# Wings of Gold

# Wings of Gold

## An Account of Naval Aviation Training in World War II

### The Correspondence of Aviation Cadet/Ensign Robert R. Rea

EDITED AND WITH INTRODUCTION BY
WESLEY PHILLIPS NEWTON
AND
ROBERT R. REA

The University of Alabama Press

Tuscaloosa and London

The publication of this book has been assisted by a grant
from the Auburn University Humanities Fund.

LIBRARY OF CONGRESS CATALOGING-IN-PUBLICATION DATA

Rea, Robert Right, 1922–
  Wings of gold.

  Bibliography: p.
  Includes index.
  1. Rea, Robert Right, 1922–     —Correspondence.
2. World War, 1939–1945—Aerial operations, American.
3. World War, 1939–1945—Personal narratives, American.
4. World War, 1939–1945—Naval operations, American.
5. Aeronautics, Military—Study and teaching—United
States.   6. Fighter pilots—United States—Biography.
7. United States.   Navy—Aviation—Biography.
I. Newton, Wesley Phillips.   II. Title.
D790.R38   1987     940.54'4973     86-7013
ISBN 0-8173-0319-7

*For*
*Merlin Owen Newton*
*and*
*Phyllis Edwards Rea*

# CONTENTS

# ILLUSTRATIONS

# PREFACE

Forty years ago, the United States emerged victorious from the greatest conflict in which the nations of the world had ever been engaged. In the succeeding years, World War II has been dissected and reassembled in every conceivable manner. Statesmen, generals, and admirals have explained their roles and described the global course of the war; the lesser ranks who fought it on land and sea and in the air have recalled its impact at the most fundamental level. Its enduring historical interest is attested by a swelling procession of books, scholarly and popular; of movies, new and endlessly rerun; of television series that seem to last as long as the war itself. Each succeeding generation of young Americans has been touched by the drama and subjected to its revisions. Appropriately so—for even more than its predecessor in 1914–1918, World War II set the world on a new course, jarred civilization out of its established orbit, made the second half of the twentieth century a new era in human history.

It might be argued that as its final achievement the war harnessed men to machines as never before in their experience, offering as its legacy the threat of nuclear annihilation. Human valor and human suffering there were, but technology was the key to victory. The impact of technology has been extensively recognized, particularly in recent years as more and more scientific and military secrets have emerged from the memoirs and archives. Less visible, because less exciting or less clear in historical form, are the paths by which men, machines, and weapons were fused into the abstractions called armies, navies, and air forces, of which great men wrote, in which lesser men were lost to sight. Yet without proper training, men could not wield the deadly weapons nor master the death-dealing machines. The significance of training is implicit in every military history, yet relatively little attention has been given to it as a specific aspect of the war. Apart from uniformed service, the most basic experience that soldiers, sailors, and airmen shared was training, for a minority of Americans under arms saw combat. Nowhere was training more critical to the final outcome of the war than in the air.

This book presents one aspect of the history of preparing men for war, specifically the training of a U.S. Naval Aviator, one of nearly 30,000 who suddenly found themselves part of a very young branch of a proud old service. They were civilians to be fitted into a martial mold, earthmen to be transformed into airmen and distinguished from all others by their wings of gold. There is, in this personal account of pilot training, nothing of heroism and much of boredom, routine, red tape, the normal accompaniment of Everyman's war—which is why training is so frequently ignored, even though it reflects the universal experience.

The project grew from Wesley Phillips Newton's scholarly interest in twentieth-century aviation. His investigation of World War II military flying led to conversations with his colleague, Robert R. Rea, and from these emerged the idea of describing the wartime flight training process in the U.S. Navy through the letters that have been printed here. As historians, the authors recognize that the experiences related were unique, referring as they do to a single individual, but the system was much the same for all—or must be assumed to have been so until many more personal accounts have been brought together for comparison. The project disclosed a surprising absence of published information about the system itself. Not only do official histories neglect the whole training process, but the unpublished histories are chiefly concerned with the material and administrative aspects of a constantly changing scene, with plans and command structures, and with what was supposed to happen rather than what did happen. The latter concern is quite rightly that of historians and is the primary justification of this book.

The authors are happy to acknowledge their obligations to the following: Bonnie Knauss and Kennette J. Harder of the William Jewell College library, and Pat Jones at the University of Iowa library, for material bearing upon the Naval Flight Preparatory School and Preflight units at their respective institutions; Dean Allard, Naval Historical Center, Operational Archives, Washington, D.C.; William J. Armstrong, U.S. Naval Air Systems Command, Washington, D.C.; Roy A. Grossnick, Naval Aviation History and Archives, Washington, D.C.; and Clark G. Reynolds, Patriots Point Naval and Maritime Museum, Charleston, South Carolina; David A. Long, Naval Historical Foundation, Washington, D.C.; Clayton R. Barrow, Jr., and Patricia M. Maddocks, United States Naval Institute, Annapolis, Maryland; Nicholas A. Komons, Federal Aviation Administration, Washington, D.C.; Helen Hopewell, Air University Library, Maxwell AFB, Montgomery, Alabama; and T. L. Reese, Public Affairs Officer, NAS Corpus Christi, Texas, for aid and advice, photographs, and materials that were invaluable in preparing the introduction on the history of training naval aviators and illustrating the text; and Auburn University and the Auburn University Humanities Fund for released time and financial support. We have sought to protect the privacy of old friends and fellow cadets, and the Navy's adherence to alphabetical groupings should provide them with anonymity from any but themselves. A few must accept the fact that they were heroes in one cadet's eyes. To those Naval Aviators who led the way and did the job, we are everlastingly grateful.

Auburn University                    ROBERT R. REA
                                     WESLEY PHILLIPS NEWTON

# Wings of Gold

# Introduction:
# The Training of U.S. Naval
# Aviators, 1910–1945

THE BIRTH OF U.S. Naval Aviation was the result of the foresight and daring of Captain Washington Irving Chambers, USN, and the civilians Eugene Ely and Glenn H. Curtiss. Chambers, a "battleship sailor," was an engineer with an eye to the future; Ely demonstrated the feasibility of seaborne operations by making the first takeoffs and landings aboard the makeshift flight decks of 1910 and 1911. Curtiss, hoping for future sales, volunteered to train a naval officer as an aviator. The first fledgling, Lieutenant Theodore G. "Spuds" Ellyson, transferred from one hazardous and only slightly older duty, submarines, to the newest. Reporting at the end of the year 1910, he became one of the first pupils at Curtiss's new training and experimental facility on North Island, a barren strip of sand off San Diego, California. Ellyson also advised Chambers of Curtiss's success in developing a "machine able to rise from the water that is smooth at sea and that can be hoisted out and in like a boat."[1] In February 1911, Curtiss demonstrated that his hydroplane could be hoisted on and off a warship not designed to handle planes. The first Navy Department appropriation for aviation made possible the purchase of a Curtiss hydroplane, the *Triad,* which Ellyson eventually flew.

At the end of March 1911, Ellyson reported to Chambers on the progress of his training in Curtiss's four- and eight-cylinder pusher landplanes. "March 4th, made four flights over the half mile course." "March 6th. Made six flights over the long course . . . height twenty feet." These were apparently straight-ahead solo flights; in any new phase of instruction, Curtiss took the controls. "March 16th. Made two flights with Mr. Curtiss, in order to get the feel of the machine in turning before attempting to turn myself." Then, on March 18: "Made two trips over the mile and a half course, and three trips over a circular course of approximately five miles. This was the first time that I had been allowed to make a turn." The primitive technology and the inherent dangers of flight were illustrated by Ellyson's account of a flight on March 18: "The radiator boiled and the water which overflowed grounded the magneto and the engine cut out. At this time I was at a height of about fifty feet and over a level spot, so was able to land safely and without damage to the machine."[2]

Curtiss's teaching methods were practical: "March 23d. Mr. Curtiss made a flight using only one aileron, in order to demonstrate the fact that in case one aileron should carry away the machine could still be easily controlled."[3] The Navy's first student aviator held the first instructor of naval flight personnel in the highest esteem. "It was not Curtiss, the genius and inventor," whom Ellyson remembered, but

> "G.H.," a comrade and chum, who made us feel that we were all working together, and that our ideas and advice were really of some value. It was never a case of "do this," or "do that," . . . but always, "What do you think of making this change?" He was always willing to listen to any argument but generally managed to convince you that his plan was best.

Ellyson graduated from the Curtiss school on April 11, 1911, and Curtiss informed the Navy Department, "Lieutenant Ellyson is now competent to care for and operate Curtiss aeroplanes and instruct others in the operation of these machines."[4]

The next Navy flight trainee, Lieutenant John Rodgers, took to the air as part of an agreement between the Wright brothers and the Navy Department whereby the Wrights provided free instruction at Dayton, Ohio, and the Navy purchased a Wright pusher. Rodgers began his training in March 1911 and after a few months became certified as Naval Aviator No. 2. The third Navy pilot, Lieutenant (J.G.) John H. Towers, was trained by Ellyson at Hammondsport, New York, where Ellyson and Curtiss were testing the *Triad* in the summer of 1911. In September 1911, the first Naval Aviation base was opened at Annapolis, and there Lieutenant (J.G.) Patrick N. L. Bellinger and Ensign V. D. Herbster received flight training. In the next two years, at Annapolis, at North Island, and at the headquarters of several aircraft manufacturers, a dozen Navy and Marine officers received flight training.[5]

The year 1914 witnessed major advancements in Naval Aviation. An Aeronautic Center was established at Pensacola, Florida, where flight and ground training would be focused. Hardly had the Center been launched when an international confrontation erupted in Mexico and President Woodrow Wilson ordered the Navy to blockade Tampico. In April, the Navy Department ordered Lieutenant Commander Henry C. Mustin, in charge of the Center at Pensacola, to send aviation detachments to Mexico. Mustin, Naval Aviator No. 11, led the unit that arrived at Veracruz, and on April 25, Lieutenant Bellinger, Naval Aviator No. 8, piloted a Curtiss flying boat over the harbor and town on a reconnaissance sortie, the first official combat flight of a U.S. aircraft. Although they were occasionally fired upon by the Mexicans, the Navy pilots, flying only at Veracruz, never retaliated while carrying out their observation missions. The Mexican experience was valuable, but the Navy flyers were happy to return to Pensacola in June to resume their training and experimental routines.[6]

No flight crew fatalities occurred during the Mexican operations, but Naval Aviation had already lost Ensign W. D. Billingsley, the first student pilot to be killed, in June 1913, and Lieutenant (J.G.) J. M. Murray, Naval Aviator No. 10, the first Pensacola casualty, in February 1914. In view of the primitive state of technology and aerology, it was probably a testimony to their seat-of-the-pants flying skills that the mortality rate among early Navy flyers was so low.[7] Accidents did contribute to phasing out pusher-type aircraft in 1915 in favor of tractor designs, with their engines mounted at the front of the aircraft.[8] By then, however, European aviation technology had surpassed that of the United States, and as World War I unfolded, the U.S. Army and Navy were left far behind.

A few signs of progress in Naval Aviation appeared in 1916. Secretary of the Navy Josephus Daniels authorized a special uniform and wings insignia for Navy flyers, and qualifications were established for an Aeronautical Force in the Naval Militia. The Aeronautical Force had few aircraft to work with, but a small group of enlisted men began flight training at Pensacola, the first who were not graduates of Annapolis. At the end of 1916, the Navy had only twenty-six regular aviators and thirty trainees.[9]

Never a decisive weapon in World War I, the airplane was used primarily as an adjunct to land warfare. Hydroplanes and flying boats were, however, employed by both sides to attack enemy surface vessels when they were within operational range of aircraft bases, and naval aircraft also became important for the Allies as an antisubmarine weapon. In 1915, a British seaplane released an aerial torpedo that sank a Turkish merchant ship in the first and only successful use of this weapon against a vessel at sea during the war. During the Battle of Jutland, in 1916, a British seaplane made the pioneer reconnaissance flight during a surface engagement. The British also refitted several vessels as crude carriers. Planes flying from these ships shot down a German Zeppelin and bombed a Zeppelin base.[10]

When the United States entered the war in 1917, the air arm of the U.S. Navy was woefully unprepared for combat, but eventually naval aviation made a respectable contribution to the war effort. Allied naval air patrols, including aircraft of the U.S. Navy operating from European and American bases, gave vital protection to transatlantic convoys at both ends of their journeys. In Europe, the Navy organized the Northern Bombing Group, composed of Marine and Navy units, for the bombing of German naval bases and inland targets, but a scarcity of planes thwarted their operations. Individual pilots of the Group were given permission to fly with the British or French on the Western Front, and a number of American naval aviators joined in the final air battles of the war.[11]

The production of the air crew, mechanics, and support personnel for these various operations required money, planning, time, new bases, and new equipment. There was only a small nucleus to build upon in April, 1917: some forty officers of whom thirty-nine were aviators, fifty-four planes of which none was combat worthy, three balloons, one airship that was not in condition to fly, and one base. Although the Army's air arm would have funding priority, Naval Aviation achieved a severalfold expansion. When Congress declared war in April 1917, Pensacola, the sole Navy air base, had a capacity for training only sixty-four aviators and an equal number of mechanics. Before many months passed, primary training bases sprang up at East Greenwich, Rhode Island; Miami and Key West, Florida; and San Diego. Curtiss provided advanced pilot training, which was directed by pioneer aviator P. N. L. Bellinger, at the company's school at Newport News, Virginia. Naval Militia stations at Bay Shore, New York, and in Squantum, Massachusetts became training bases, while the Goodrich plant at Akron, Ohio, hosted a school for balloon and blimp instruction. By the end of the war, Bay Shore, Miami, and Key West had become the chief primary bases; Pensacola and Hampton Beach, near Newport News, the principal advanced training bases; and the first ground schools had opened at the Massachusetts Institute of Technology, the University of Washington, and Dunwoody Institute. Aviation students also began to study the complexities of aerology and radio at Harvard University. Most wartime flyers were enlisted, trained, and commissioned in the Naval Reserve Flying Corps. Along with a few old Militia men, a group of Yale University students and air enthusiasts formed the nucleus of the Reserve.[12]

Pensacola Naval Air Station was both a primary and an advanced training base until May 1918. The pressing need for aviators led to "a training program that operated from dawn to dusk seven days a week. After dark, a night shift repaired planes in the newly constructed, electrically lighted hangars." The eighteen-month-long training program consisted of increments devoted to aerial gunnery, bombing, navigation, signaling, photography, radio, aircraft rigging, and nomenclature of planes and engines. In the primary flight phase, students usually soloed in the Curtiss N-9 "seaplane Jenny," a tractor-type biplane with pontoons. Ten hours of flying seems to have been considered the minimum to qualify a pilot. The regimen included stalls, loops, and flying over water. Extended coastal reconnaissance missions were flown in the Curtiss R-type seaplane. A "typical navigational problem was to return to the point of departure after flying an equilateral triangle course, forty miles to the leg." Students received gunnery instruction on a deserted strip of Santa Rosa Island, where they fired "at fixed and moving targets with a machine gun mounted in the front seat of the plane." For bombing practice students employed bricks and dummy bombs before being given live ammunition.[13]

Other training included catapult launching from a coal barge and signaling with a manually operated blinker device. It was worth the sweat of humid summer days to fly over the Gulf of Mexico in the open cockpit of a Curtiss seaplane, clothed in a tan sheepskin long coat, a moleskin hood, goggles, black leather gloves, soft leather boots, and a lifebelt.[14] The wind whipped against the air screen as the plane bounced up and down and skidded a bit to either side; the drone of the OX engine hammered against the eardrums; below, the variegated surface of the Gulf shimmered in hues of purple, blue, and green, while occasional swells added brushstrokes of white foam.

Accidental deaths in training took their toll at Pensacola. Twenty-two students died in air accidents, nineteen of them in the last nine months of the war, a clear indication that training had intensified.[15]

After the armistice on November 11, 1918, the Navy and its air arm were subjected to severe reductions. Budgets and bases were limited, but by 1922, the collier *Jupiter* had been converted into the aircraft carrier *Langley,* and she would be joined by the converted carriers *Lexington* and *Saratoga* in 1928. Heading the drive for a carrier-based Navy was Rear Admiral William A. "Billy" Moffett. He had to contend with another "Billy," Brigadier General William L. Mitchell of the Army Air Service, whose abrasive efforts to achieve an independent air force controlling both sea- and land-oriented aviation probably spurred the Navy to create the Bureau of Aeronautics.[16]

Mitchell overstepped himself with his scathing criticism of both Navy and War departments, and the latter clipped his wings at a court-martial in 1925. Moffett, on the other hand, continued to head Naval Aviation and advance the cause of the aircraft carrier until his death in April 1933, when he was killed in the crash of the airship, USS *Akron.*[17]

The 1920s saw significant advances in technology. The Navy experimented with the aerial torpedo, whose use in combat had been pioneered by the British. By 1923, Navy tests had revealed that "torpedoes could be successfully launched from aircraft and be made to run straight."[18] Divebombing also won attention. In 1927, U.S. Marines found themselves in the thick of fighting in Latin America during an American intervention in Nicaragua. When Nicaraguan guerrillas cut off a detachment of leathernecks in the village of Ocotal, a Marine air unit made the first organized divebombing attack and scattered the partisans of a *jefe* whose name became legendary—Augusto Sandino.[19] Development of divebombing by the Navy and its performance as a stunt by exhibition flyers in the United States influenced the Germans to incorporate it as a basic tactic of the new Luftwaffe.

The first U.S. aircraft carrier, *Langley,* became a floating laboratory for divebombers and other operational planes and for working out carrier procedures and safety rules. In 1929, *Lexington* and *Saratoga* engaged in fleet exercises off the Panama Canal Zone, and *Saratoga*'s planes staged a mock attack on the Canal which was judged by referees to have destroyed the locks and blocked the interoceanic artery. It was a triumphal day for carrier warfare and particularly impressed the next Chief of Naval Operations, Admiral William W. Pratt, who saw to it that carriers formed the heart of an independent offensive force in the fleet.[20]

New ships and planes required new pilots, but the sharp cutbacks in the Navy's air arm after World War I created a bleak outlook for the training of young aviators. Pensacola was still the main training base, but a shortage of qualified candidates limited its productivity to twenty trainees per class in 1919, and the length of the program was reduced in 1920 for lack of funding. The numbers of graduates and of those who failed to complete the course tell a dismal story. Only 91 men qualified as aviators in 1922. A third of the 170 who began training in 1923 failed to qualify. The high failure rate resulted from the return to the prewar practice of limiting officer trainees to Naval Academy graduates who requested aviation training and who had served two years at sea. By the time these men reached Pensacola, many had lost the sharp edge of youth demanded by the training regimen. The Navy Department therefore decided to augment the program with enlisted trainees, although they would not be commissioned upon qualification but would be designated Naval Aviation Pilots, whereas officer graduates became Naval Aviators. Aviation Pilots were further restricted to training, patrol, torpedo, ferry, and utility duty.[21]

The Navy Department inexplicably worsened the situation between 1922 and 1923 by directing that officer graduates of flight training should be assigned to nonflying sea duty for one year prior to being designated Naval Aviators. Throughout the Navy, junior officers began to resign in large numbers. Training accidents also increased in 1923 and 1924, causing seven fatalities at Pensacola. The accidental deaths of Naval Aviators Nos. 1 and 2, Theodore Ellyson and John Rodgers, could not have helped morale. In an effort to reduce the restraining influence of anxious families upon young officers, trainees were required to live on base in bachelor quarters. Base authorities prepared for the worst by stationing ambulances where training flights were taking place and tried to forestall accidents by requiring before each flight that both instructor and student sign statements that they had not consumed alcoholic beverages in the previous twenty-four hours. The evidence does not prove or disprove the effectiveness of this station regulation.[22]

When a student reported for aviation training at Pensacola in the 1920s, he came under the command of the superintendent of flight training and the senior flight instructor and was subject to the observation of flight surgeons who conducted thorough physical and psychological examinations before training commenced. Each candidate was assigned to a flight in one of the two training wings of his class. Besides helmet, goggles, and other personal equipment, students received a copy of the *Flight School Manual*. Its preface informed the neophyte:

Flying cannot be learned out of a book. Only time in the air will make a pilot, and the time required to produce a thoroughly competent and experienced pilot runs into hundreds of hours. But a student who, before getting into a plane, has a reasonably clear idea of what he is attempting to learn, will derive considerably greater profit from time in the air than one who has not.[23]

Most instructors were officers who had been trained for World War I and were among the few who had chosen and been able to remain on active duty. Gradually, with expansion and the integration of aviation into the fleet, the instructors were veterans of fleet air squadrons. Each instructor worked with four students, two from each of the training wings. During a given week, one wing engaged in flight training in the morning while the other attended ground school, and in the following week they alternated times for flight and ground training. By 1927, four classes a year were being graduated from Pensacola.[24]

In the sedate old Curtiss N-9 seaplane that introduced students to primary training from 1917 to 1927, the student sat in the rear cockpit, wearing a Gosport flying helmet "with earphones and a voice tube leading to a mouthpiece used by the instructor," in what was strictly one-way communication, while the neophyte worked his set of dual controls. During the first ten hours of dual instruction, the student tried to keep the plane straight and level while maintaining sufficient speed to keep it from stalling. He then practiced turning and banking, seeking the "proper coordination in the use of the rudder and the aileron" in order to effect "reasonably smooth turns with a fairly constant bank" to avoid slips and skids. During these first ten hours, much time was devoted to landing, for it was the most difficult maneuver. The full stall landing was used exclusively at this point: the plane was "allowed to lose flying speed and drop into the water from a height of one to two feet."

At the end of the first ten hours came the initial hurdle, the first solo flight, unless the student needed three more hours of practice. Those who qualified for the solo at the end of the first ten or thirteen hours did so as the result of a positive rating from at least two of the three persons with whom the student had taken check rides—his instructor, the chief (senior) instructor, and the assistant chief. Because the selection process began here, "rumor and speculation" were "rife." Students who failed the solo check faced a board consisting of the officer in charge of flight training and the check pilots who had failed them.

The result was usually dismissal from the program and a hasty departure from the Pensacola railway station.

For those students who qualified for "once around the short course," there was the thrill of a five-minute solo flight. In solo flights that followed, they found that "the extreme emptiness of the forward cockpit is hard to forget. . . . But . . . by the end of five hours the average student is steadily building up confidence in himself." In the course of the next twenty-five hours after soloing, a student received instruction in such maneuvers as spirals "in which the plane is banked more than forty-five degrees, so that the rudder acts principally in a vertical plane and the elevators in a horizontal plane," loops, split S, and falling leaf. At the end of each five hours, a check pilot rode with a student to observe his progress. Landing was made more difficult in a maneuver that began at 6,000 feet with a power spiral to 3,000 feet, then required cutting the engine switch and gliding, turning twice more in a spiral, and bringing the plane down within a 200-foot radius of a set mark.

A beginning student had to exhibit an "irreducible minimum" of aptitude for flying, but too much aptitude sometimes bred overconfidence and carelessness. As a senior officer warned, "The student who, after starting his engine on the gravity tank, takes gaily off without turning on the main tank is not going to get very far. Yet in every class two or three students have to make forced landings before this fact impresses itself upon them." A student had to be alert: "Frequent exercise of the muscles of the neck is one of the surest guarantees of a ripe old age"—sage advice later reduced to "Better a stiff neck than a broken neck." A student had to be accurate "in detail of performance." Landing even a few degrees off wind in seaplanes was "considered sloppy flying. . . . If a student can land at all, he can just as well land directly into the wind."

After a few hours of formation flying in three-plane vees, a bit of night flying, and some more sophisticated maneuvers such as the wingover and the eight, came the final check in the N-9, during which the chief instructor or his assistant looked for "accuracy and precision . . . in take offs, turns, spirals, and landings." Students had to perform the wingover and eight without stalling and "without violent slips or skids." Training in more advanced seaplanes was abandoned in 1926, and students went directly from the basic seaplane trainer to landplanes.[25]

All operational Navy planes of the 1920s were fabric-covered biplanes, as was the standard Navy landplane trainer, the Consolidated NY-1, adopted in 1925. Landplane training took place at Corry Field, several miles north of Pensacola. The NY-1 was light and sensitive to the touch; the biggest problem was learning to land it, for both the approach pattern and landing space were more restricted than they had been on Pensacola Bay. To acquaint a student with a limited landing space and emergency landing procedures, the instructor would suddenly close the throttle. If a student expected to get down safely he had to select a field within gliding distance. The instructor waited as long as possible "to make the theoretical result of the student's judgment evident," then he reopened the throttle. The successful student eventually became proficient in "shooting" fields with a simulated "dead stick," and "any instructor who is troubled with ennui, experiences a marked re-awakening of interest at this point."

Students next received bombing and torpedo training in the operational Curtiss CS-1, and the final phase of instruction stressed aerial gunnery, bombing, and navigation. While flight training progressed, students also underwent classroom and practical instruction in aircraft construction, engine maintenance, and radio repairs. The Navy was beginning to take an interest in aerology, sending officers to Harvard or M.I.T. for basic instruction, then to the Weather Bureau for more advanced training.[26]

By 1927, a basic program with established techniques had been developed and would persist in the face of changing technology. Its philosophy, however, would undergo considerable alteration. Writing at that time, Lieutenant Barrett Studley, a Navy flight instructor, stressed that "flying is not a science. It is an art." He was, like many another flyer of the day, civilian or military, a devotee of the seat-of-the-pants approach to flying. Lieutenant Commander DeWitt C. Ramsey, writing in 1927 from a desk in the Bureau of Aeronautics, but on the basis of his on-the-spot observations, not only characterized flying as an "art" but also lauded training planes whose design was a decade old. He seems to have assumed that a crash was inevitable, "due to human rather than to material frailties," but "provided physical members can be readjusted and nerves are not too badly shaken, an aviator emerges from the wreckage of his initial crash, a sadder, but a much wiser and more useful pilot."[27]

The crashes of Navy planes were, as a matter of fact, quite frequent, and the fatalities, for peacetime, were alarming. There was an average of 15 fatal accidents each year from 1920 to 1929, and 233 deaths in training and operational flying. Given the seat-of-the-pants philosophy, the backwardness of aerology, and the state of technology, these figures are not surprising. The washout rate at Pensacola, in the 1920s, averaged about 50 percent, but it was far higher in some years. In 1929, for example, 63 officers pinned on their gold wings; 84 did not. Twenty-seven enlisted men became Aviation Pilots; 236 washed out. Most who did qualify in those years, with 230 hours of flying in six months' time, went into fleet aviation squadrons, where unit indoctrination and polishing took place.[28] Such was the background of the men who flew off *Saratoga* in 1929 and theoretically destroyed the Panama Canal.

Admiral Moffett worked diligently to advance a Reserve program after World War I. Arguing that war veterans needed to hone their skills and newcomers must be attracted if a Reserve Flying Corps were to be in a proper state of readiness, he managed to gain fifteen days' annual training for inactive reservists in 1922 and the establishment of one Reserve flying unit in each Naval District. In 1923 he arranged to give forty-five days' preliminary flight training, followed by a similar period at a Naval Air Station. Those who completed both phases and passed an examination might receive a commission in the Naval Reserve and wings of gold. Funding was sporadic, and Moffett was forced to call upon private military academies and the civilian flying clubs that were sprouting up in that day of the Winged Gospel for training facilities.[29] Both commissioned and enlisted reservists might receive further instruction at Pensacola if they agreed to serve a year with the fleet.

In 1925, the first postwar Naval Reserve Air Bases (NRAB) were established at Seattle and Boston. Their mission was to provide basic flight training and to boost the efficiency of experienced pilots, but they were woefully short of both men and planes. The NRAB at Long Beach, California, was commissioned in April 1928, but not until June did it receive its first plane, an old Vought UO-1. The ancient trainer was "rather heavy, with a weak landing gear spreader bar that would let go if a landing was made in a side slip," yet it was kept as "busy as any airplane could be on a gas-available basis." The first enlisted reservists were sworn in and given ground training in July, but until the following year it was necessary to transfer them to the better-equipped NRAB at Seattle for flight training.[30]

In 1927, Lieutenant Commander D. C. Ramsey at BuAer pointed to other aspects of aviation training that deserved attention. One was the requirement that all officers at or above the rank of lieutenant commander take flight training at Pensacola. Eventually, either by order or by request, several older officers received that training, and their number included such future luminaries as Ernest J. King and William F.

Halsey, Jr. There was also an effort to make aviation training an integral part of the program at the Naval Academy—once the exclusive source of trainees and the site of the first Naval Air Base. One-third of the members of the first class at Annapolis did not participate in the annual summer cruise, and in 1925 these cadets received indoctrination with fleet air squadrons. In the summers of 1926 through 1928, all new ensigns from the Academy received this same indoctrination.

In 1929, the Navy Department assigned a squadron, VN-8, to provide indoctrination flights to Annapolis midshipmen. They were given courses directly related to aeronautics in five different departments, including such subjects as theoretical bombing, aviation gunnery, and aerology. Squadron 8 accomplished one of its major goals. After 1930, there was a sharp increase in the number of midshipmen who wanted to become Naval Aviators.[31]

In 1930, aviation squadrons in the fleet were divided among the Battle Fleet, the Scouting Fleet, and the Asiatic Fleet. Included in the Battle Fleet were the carriers *Langley, Lexington,* and *Saratoga,* as a tactical unit, with their bombing (VB), fighting (VF), torpedo (VT), and scouting (VS) squadrons. Among them, the three carriers held 164 out of a total of 875 operational Navy planes. In the fleet and billetted in a growing number of shore stations, there were 614 Naval Aviators and 244 Naval Aviation Pilots at midyear, 1930.[32]

The flight training syllabus in use in 1930 showed little change in techniques and concepts of instruction. With a total of 227.75 hours of air time for officers and 167.75 for enlisted men, it included mandatory night flying but no practice in bombing, torpedo work, or aerial gunnery. Ground school alone provided instruction in the theory of torpedoes and bombing. Some of the course descriptions suggest the content of ground instruction at this time. "*Aerodynamics*:—General consideration of airfoils and applications of the wind tunnel; parasite resistance; stability; controllability; maneuverability; the propeller; performance and dynamic loads." "*Range gunnery.*—Sixteen practices with screen and model targets, stationary and moving, using Lewis and Browning guns, the latter being mounted in rocking nacelle; trap shooting." In 1931, a revised syllabus restored courses in aerial gunnery and bombing and torpedo training and extended flight time to 282.75 hours for officer trainees and 258.75 for enlisted men.[33]

By 1933, the depression notwithstanding, Naval Aviation began to move forward. President Franklin D. Roosevelt authorized funds for new Navy ships, including the two aircraft carriers *Yorktown* and *Enter-*

*prise*. In 1934, *Ranger* was commissioned, the first American carrier designed and built as a carrier from the keel up, and Congress approved another carrier, *Wasp*. By the mid-1930s, aircraft technology was being transformed by such features as stressed metal skin, internally braced wings, retractable landing gear, supercharged engines, controllable-pitch propellers, and improved radio communications. To launch improved aircraft such as the Douglas Dauntless divebomber from carrier decks and to check their roll once they had landed, engineers perfected hydraulic catapults and arresting gear.[34]

Although aviation became well integrated in the fleet during the 1930s, there was still a chronic shortage of both men and planes. To provide more pilots, the Navy authorized additional instruction for students who washed out or interrupted their training for other reasons. To encourage both commissioned and noncommissioned pilots to return to flying after long stints of nonflying duties, a requalification course was introduced.[35]

The year 1935 saw a profound change as the Navy adopted the cadet training system which the Army's air arm had long utilized. Public Act 37, the Naval Aviation Cadet Act, established the grade of Aviation Cadet in both the Navy and Marine Corps Reserve. The act provided for one year of pilot training for qualified college graduates between the ages of eighteen and twenty-eight, with pay, uniform allowances, and insurance. Upon completion of training, they were required to serve on active duty for three years as ensigns or second lieutenants; then they would be granted a bonus of $1,500 and would assume inactive duty status in the Reserve. The act was a result of lobbying by Admiral Moffett and by his successor as Chief of the Bureau of Aeronautics, Rear Admiral Ernest J. King, which persuaded the administration that it would provide employment for some bright young men in line with the expansion of the Navy and Naval Aviation.[36]

For the first cadet class, which began training in July 1935, there was a new training syllabus. It entailed about 300 hours of flight training and 465 hours of ground instruction. While the first cadet classes received some primary training in the old Consolidated NYs, it was not long before the two trainers known as "Yellow Perils" made their debut. In December 1934, the Stearman biplane, to be designated NS, received its final testing; and in March 1936, the Naval Aircraft Factory's N3N biplane trainer went through its final checkout. The color of these planes rather than their flying characteristics led to the nickname cherished by generations of cadets.[37]

The pre–World War II cadet program has been vividly described by Captain Joe Hill, USN (Ret.). A native of Indiana, Hill encountered physical standards requiring a height ranging from five feet six inches to six feet four inches, weight to match height, possession of most of the teeth and these in good condition, 20/20 vision and precise depth perception, and acute hearing. No beards or mustaches were allowed.

Either applicants for the cadet program went to a Naval Reserve Air Base, or a recruiting team visited them on a college campus, where physical and written exams were administered. Once accepted, the potential flyer underwent thirty days of precadet training as a seaman second class-V5 at a Naval Reserve Air Base. Because the elimination rate was greater than 30 percent, these were known as E bases. In addition to ten hours of dual flight instruction at the E bases, the S2c's had many hours of ground training in naval terminology, parachute packing, engines, and flight instruments, but mostly, according to Hill, "they learned about washing down airplanes and cleaning heads (toilets)."

The final exam in the precadet phase was a solo flight. Once passed, the fledglings were ready to sign a contract to become Aviation Cadets and to receive one year's training at the pay rate of $75 per month with room and board. When training had been successfully completed, cadets were commissioned and designated Naval Aviators, with initial pay of $125 a month, plus a $30 food allowance. They were given quarters aboard ship while they were on duty with the fleet or had bachelor officers' quarters ashore. The new Naval Aviators were required to serve two years with the fleet and a final year on active duty at a Naval Reserve Air Base. They had to agree to remain unmarried for the full four years of the contract. At the end of this time they reverted to inactive Reserve status and received a bonus of $500 for each year they had served as a Naval Aviator.

Hill described a number of cadets who made it through training in the classes of 1935–1940. It is interesting to compare their backgrounds, motivations, and experiences with those of the wartime generation of Robert Rea, whose story follows.

Because most prewar Naval Aviation Cadets held bachelor's degrees or had senior status in college, they usually came from middle- rather than working-class backgrounds, but these were the years of the Great Depression, when money was scarce, and many including Hill himself, had had to work in order to meet their college expenses.

Motivation was as varied as the men themselves. Cadets like Walter C. Michaels, Ben Troemel, John L. DeLorenzo, and Joe Hill gave Charles

A. Lindbergh credit for having sparked their interest in flight. Ed Kiem's curiosity had been aroused by the humorist Will Rogers, a great booster of flying who died in an air accident. As boys, Hill, David M. Burns, Jack L. Poteat, Harvey Hop, and Al Dodson watched the feats of that disappearing breed the barnstormers and were inspired to fly. Hop had the added thrill of taking a flight with the celebrated aviator Clarence Chamberlain, one of the first transatlantic flyers. Hop and DeLorenzo built model planes. Howard Dostal flew gliders with a Boy Scout troop and later with a club at Purdue University, where Art Farwell also soared. Charles Ingalls, who studied aeronautical engineering at Auburn (Alabama Polytechnic Institute), first tried to become an Army Aviation Cadet but was turned down because he had severe acne; he persisted, became a Naval Aviation Cadet, and won his commission. A Navy recruiting film entitled *Wings of Gold* helped to lure DeLorenzo, who liked the looks of the F4B and the girls on the beaches at Pensacola. David Burns was among the men who shared a tent and slept on army cots at the NRAB at Opa Locka, Florida. Flying began in an old Curtiss biplane with a tail skid and no brakes; takeoffs and landings, including the up-or-out solo, were on grass.[38]

Those who survived the E bases became Aviation Cadets at a Pensacola rich in the tradition of training but unaccustomed to such an influx of trainees. Hill described their quarters:

> The barracks were well constructed two-storied brick buildings, with 4 wings, each housing up to 75 cadets. Each wing had 12 wash basins, but only 6 mirrors for shaving. There were 6 open showers, 6 urinals, and 6 heads. . . . The bunks were metal double deckers with good sized mattresses, and each cadet had a spacious locker and some shelving. Each wing had a large screened porch, which was ideal for early morning calisthenics. The entire ground floor of a central building was used as a cadet mess. . . . The food, by most standards, was excellent and was served in good style.[39]

The base also provided a library, nightly movies, bowling, pool or billiard facilities, and fields for outdoor sports. Hill recalled that it cost two dollars to join the Officers' Club, and "you could have a large evening with a date on Friday or Saturday night for two or three dollars." Upon checking in, a cadet was measured and furnished with one blue dress uniform, two white dress uniforms, and four pairs of shoes, one black, one white, and two brown, the latter being worn in the Navy only by Aviation Cadets or aviators.

Barrett Studley, now a Lieutenant Commander, was the first officer in charge of cadet indoctrination. Studley had played a major role in preparing the "Indoctrinal Training Syllabus," and his experience as a "mustang" (an officer commissioned from enlisted status) manifested itself in this phase of training. Hill noted that "he kept them [the first cadet class] busy from 5 a.m. until 8 p.m. every working day for 6 weeks." The syllabus included naval governance, history of naval aviation, the cadet plan, the role of a Navy officer, types of vessels, fleet aviation, military etiquette in general and aboard ship, military law, basic seamanship, military drill, and the manual of arms.[40]

Studley established the standard system of discipline that had long been a part of cadet life at the service academies. The basic method of punishing such infractions as carelessness with equipment, sloppiness in clothes or quarters, or laxity in military courtesy was by awarding demerits, which had to be "walked off" on weekends by delinquent cadets, rifle on shoulder.[41]

The Cadet Battalion was divided into Wings and was commanded by cadet officers. The first Cadet Commander of the Battalion was Walter Hibbs, who completed the course as top man in his class. Cadet rank was not, however, a certain indicator of success as an officer. Hibbs attained the rank of captain before he retired to become an airline pilot. His classmate, John J. Lynch, was the only member of the first cadet class to win flag rank, retiring as a rear admiral. Only two of the thirty-five or thirty-six who won their wings in this class made a career of the Navy.[42]

The flight syllabus prior to the beginning of World War II was similar to that formulated in the 1920s. For flight and ground training, a student passed through five squadrons. Squadron 1 flew the single-pontoon version of the N3N Yellow Peril, and during a six-week period, cadets learned the basics of flying. Squadron 2 introduced cadets to primary landplanes, N3Ns at first but later the Stearman N2S. In eighteen weeks students made three-plane vee formation flights, cross-country flights, and night flights; they engaged in aerobatics that ranged from wingovers and loops to split S and Immelmann turns. Specialization began to appear in Squadron 3. For nine weeks cadets flew biplanes in advanced aerobatics, three- and nine-plane formations, and engaged in radio, navigation, and free and camera gunnery training. For nine weeks in Squadron 4, cadets were acquainted with operational seaplanes, performing torpedo and horizontal bombing and alternating as pilots, copilots, navigators, bombardiers, and radio operators.

Prospective fighter pilots had their day in Squadron 5, during a final seven weeks of relatively complex maneuvers, during which plane jousted against plane and single fighters engaged sections flying the Lufbery Circle. There was also high-altitude work at 18,000 feet, cross-country flying, divebombing, and precision landing practice in anticipation of carrier assignment.[43]

From Squadron 1 through Squadron 3, flight training alternated with ground school, made more sophisticated by new theories and advancing technology. Long hours were devoted to complex engines and aircraft construction. In gunnery, the relatively difficult technique of deflection shooting was examined. Many hours were spent learning and practicing Morse code, but only five hours were devoted to aircraft radio telephony. One of the final tests before a cadet received his wings was the Morse code sending and receiving exam, and most cadets held it in awe.[44]

During the training process there were casualties, in terms of both lives and washouts. (The expression "bust-out" seems to have been popular among cadets at Pensacola in prewar days.) Happily, the first cadet class suffered no fatal accidents. In September 1936, the first fatality occurred when a plane spun in. Casualties began to climb as the number and size of cadet classes rose. In the year 1940–1941, before wartime expansion brought a sharp increase in accidents, the number of training deaths reached twelve.[45]

In an effort to prevent accidents and lower the washout rate, the Navy gave increasing attention to the physical and psychological competence of its student pilots. With respect to physique, there was "no optimum type," but an applicant must possess reserve strength, "be it either wiry or husky," and his stature must fall within "certain limits . . . which are required by purely military considerations" and by aircraft design. To determine heart strength, doctors employed the Schneider Index, a test that consisted of "placing a uniform load on the circulation and measuring the response in terms of change in pulse rate and blood pressure." Examiners insisted upon good hearing: "Very often a failing motor will give its first warning in the form of a change in sound. These sounds as well as radio and voice sounds must be perceived in the presence of and in spite of the noise of the motor." Beyond general good health, vision was the most important qualification. It must be excellent, "perception of depth and distance . . . rapid and infallible," for the "rapidity of an aviator's changes in relation to other objects precludes deliberation." The examining flight surgeon was also responsible for evaluating a candidate's psychological stability and the likelihood that he would be able to withstand "the strain of flying under any and all conditions."[46]

A cadet faced a continual weeding process, and his future depended upon the competence of his instructors and the quality of the check pilots. Captain Hill judged them accurately when he wrote, "They were all different, but they were all human. Some were aggressive, loud, and profane, while others were passive, quiet, and firm. Some could fly better than others, and some could teach better than others, but they were all dedicated and most . . . were fair." Boards of evaluation heard evidence of poor performance and, after two "down checks," determined whether to give a student further instruction. Three hours was the normal and six hours the maximum extra time allowed before a stumbling cadet faced the Naval Air Station Commandant, a final judge who rarely showed mercy.[47]

Initially, Naval Aviation cadets were required to undergo a period of advanced training with the fleet before they were commissioned, a circumstance that placed them at a disadvantage, in terms of date of rank, relative to newly arrived ensigns from the Naval Academy. In June 1939, the Aviation Cadet Act was revised to provide for the immediate commissioning of all cadets then on active service with the fleet and subsequently for the commissioning of all cadets when they finished the course at Pensacola.[48]

In September 1939, the European conflict erupted with the invasion of Poland by the Nazis. For the next eight and a half months, the atmosphere in the United States was not seriously disturbed by the new war, even though President Franklin D. Roosevelt had sounded a rallying cry against fascism some years earlier. Roosevelt had long been aware of the military potential of air power. In 1938, when the naval armaments limitation treaties had gone by the board and the Japanese were vigorously engaged in constructing a powerful carrier force, the U.S. Congress had passed the Naval Expansion Act, whereby two more carriers, Hornet and Essex, would soon join the fleet, and the Navy was authorized to maintain at least 3,000 planes. After the outbreak of war in Europe, the Navy Department sought to speed pilot recruitment by reducing the length of the training program from one year to six months. A new syllabus completed the evolution from initial training in seaplanes to basic and primary instruction in landplanes. Advanced training provided specialization in patrol, utility, or observation craft, or carrier-based fighting, bombing, or torpedo planes. Ground school was correspondingly shortened by half.[49]

The transition from peacetime to war preparedness training came about relatively slowly, but by 1940, naval aviation training was showing the stress of expansion. To accommodate increased numbers of trainees, the Navy Department ordered additional training bases established at Miami and Jacksonville, Florida, and at Corpus Christi, Texas. Captain Albert C. Read, the commandant at Pensacola, foresaw problems within the administrative structure of the training system: flight training was generally under the jurisdiction of the Bureau of Navigation rather than the Bureau of Aeronautics, and the commandants of the several Naval Districts enjoyed immediate supervision of all flight training in their respective districts. He therefore recommended, in the event that the United States went to war, that all flight training and training bases be put under the direct authority of the Commanding Officer, NAS Pensacola. This consolidation, he argued, would ensure uniformity and would prevent jurisdictional conflict regarding training. The Chief of the Bureau of Aeronautics, Rear Admiral John H. Towers, opposed such centralization on the stated ground that training centers were too widely scattered for effective administration. A sensible proposal was set aside, for the Navy Department agreed with Towers.[50]

Although the administrative structure remained unchanged, in 1940 the Navy Department acted to equip combat aircraft with armor and self-sealing gas tanks. Many of the cadets who entered the classes of 1940 and 1941 would have good reason to appreciate the installation of this life-saving equipment in the new generation of planes, the Grumman Wildcat fighter and the Douglas Dauntless divebomber—both low-wing monoplanes.[51]

When Japanese Vals, Kates, and Zeros roared in over Pearl Harbor early on the morning of December 7, 1941, the physical and psychological shock to the U.S. Navy and to the nation was enormous. Fortunately, Admiral Nagumo's planes were not targeted to strike the Navy's oil reserves, and they failed to discover the carriers of the Pacific Fleet. *Lexington* and *Enterprise* were at sea, and *Saratoga* had just been overhauled at San Diego.

To face an advancing enemy in the Pacific and to protect sea-lanes over which supplies and troops must be sent to reinforce embattled Allies across the Atlantic, the U.S. Navy possessed, on December 7, only seven large carriers and one small carrier, just over 5,000 aircraft of all types, and some 40,000 aviation personnel—but few more pilots than in World War I.[52] Design genius would bring forth the great planes of the future—the Corsair, Hellcat, Avenger, and Helldiver—and the factories of the world's greatest industrial power would mass-produce them; it would be more difficult to mass-produce the pilots and crews to fly them. Wartime expansion, technological advances, and the exigencies of combat all combined to force the Navy to reconstruct its administrative, training, and recruitment systems.

In February 1942, Admiral Towers, a very recent convert to centralization, proposed to the Commander in Chief of the Fleet, Admiral Ernest J. King, that aviation training be reorganized into four distinct phases. Initially, cadets should spend several months in training centers where physical fitness and ground school subject matter would be stressed. Primary flight training would follow, conducted at extant Naval Reserve Air Bases. Cadets would then proceed to either Pensacola or Corpus Christi for Intermediate flight training, and after receiving their wings they should be given Operational training in tactics, gunnery, and carrier landing qualification wherever facilities might be available. Most significantly, Towers urged that these specialized functional commands be removed from the jurisdiction of local naval district commandants and be supervised directly by the Navy Department.

Secretary of the Navy Frank Knox approved Towers's plan; in April he authorized the first functional training component, the Air Operational Training Command, with headquarters at Jacksonville, Florida. By the fall of 1942, three more functional commands had been implemented: Air Primary Training, Air Intermediate Training, and Air Technical Training.[53]

The new commands required a new supervisory structure, and the general reorganization of the Navy Department in 1943 provided an opportunity to clarify the location of responsibility for aviation training. In June, Admiral King recommended to Secretary Knox that the new position of Deputy Chief of Naval Operations (Air) be created. The post was established in August and was filled by Rear Admiral John S. McCain. The DCNOA was authorized "to correlate and coordinate all military aspects of policy, plans, logistics of naval aviation." At the same time, an Aviation Training Division, directly responsible to the Deputy Chief, was created, and in December, the Naval Air Training Command was instituted. Headquartered at Pensacola, its Chief reported to the DCNOA and was henceforth responsible for supervising flight and ground school activities and for coordinating all aviation training. The basic change recommended in 1940 by Captain Albert Read had finally been effected but at a distinctly higher level.[54] Aviation cadets were generally unaware of the Olympian struggle that produced these ad-

ministrative adjustments, but they were directly affected by the con-
comitant changes in the training program.

The first steps were being taken even as the master plan was being
considered. In February 1942, the old NRAB E bases were transformed
into Naval Air Stations, where primary flight training would be concentrated, and the first new phase of ground training, to be known as Pre-
flight, was announced.[55] As projected, Pre-flight was intended to prepare
cadets physically and mentally for the rigors of traditional flight training.
Former Navy football coach Lieutenant Commander Tom Hamilton set
the tone, as he would set the pace, when he declared, "We would fail in
our duty if we did not prepare them . . . , hardened by all the training
it is possible to give them."[56] Flights of rhetoric proclaiming the necessity of toughening future pilots so that they could swim oceans, climb
mountains, crawl through jungles, and kill their enemy in hand-to-hand
combat disguised the fact that Pre-flight would also indoctrinate cadets
in naval history and customs, would teach military drill and seamanship, and would introduce them to such useful subjects as communications and navigation.

Prospective sites for the new operation had been investigated in
January 1942, and early in February the Navy contracted with the
University of Iowa at Iowa City for the first Pre-flight School. Others
quickly followed: the University of North Carolina at Chapel Hill, the
University of Georgia at Athens, St. Mary's College and (in 1943) Del
Monte, both in California. The facility at Iowa was commissioned on
April 15, 1942.

The tremendous expansion of manpower required by the war effort
soon ordained the creation of further early training programs. By December 1942, the Navy had decided to interpose two new steps between
the enlistment of cadets and Pre-flight School. The first of these was
called Flight Preparatory School. On December 17, the Navy Department
announced that it had contracted with twenty colleges and universities
(ultimately only seventeen) to provide cadets with fifteen weeks of
ground training that would "prepare them more fully for the later
training stages and . . . relieve them of much of the academic burden
heretofore encountered as training progressed." The designated schools
ranged from small private colleges such as William Jewell in Missouri
to the state universities of Pennsylvania and Texas. The Navy noted
happily that the NFPS network "greatly accelerates the rate at which
young enrolled cadets will be called into actual training." Indeed it did;
the first classes began their studies in January 1943.[57]

The second new phase was, in fact, an old one. In 1939, the Civil Aeronautics Administration had launched a Civilian Pilot Training program which utilized vocational education funds to train young pilots and support the nation's commercial flying schools. By June 1942, CPT had exposed thousands of students to flight training and had moved far toward aligning its syllabus with those of the Army and Navy. As all who completed the CPT course were now required to enlist in one of the military flying services and 30 percent of all students entering naval aviation training were coming out of the civilian program, CPT provided an invaluable pool of pilots and instructors. In November 1942, the resources of the CPT system were effectively absorbed by the armed forces. The name was changed to War Training Service in December, and the Navy became directly involved in its operations on ninety-two college campuses. From 1943 until May 1944, successive classes of cadets were sent to WTS schools as they completed their courses at the new Flight Prep Schools.[58]

By August 1943, the aviation training program was stretched to its limits and had a backlog of 23,000 men waiting to begin training. In order to use some of this available manpower, the Navy began summoning recruits to active duty as ground crewmen at Naval Air Stations. Called Tarmac duty and modeled on a Canadian program of the same name, this scheme provided useful labor and practical experience on the flight line, and it increased the Navy's control over the timing and rate of cadets' entry into the training pipeline.[59]

Before America went to war, the pilot training program was producing about 2,500 pilots a year. When the Navy projected its wartime needs, early in 1942, the planners proposed to train 30,000 pilots a year. There was no shortage of volunteers, but the Navy's physical and educational standards were demanding, and there was competition for good men from the Army Air Forces (AAF) and from draft-exempt industries. To identify, recruit, and hold the men it needed, the Navy instituted a complex series of programs that allowed recruits to pursue their civilian education until they were ordered to active duty.

The prewar V-5 Naval Reserve program formed the heart of the manpower control system. Open to third- and fourth-year college students, it allowed them to complete the current school year before being called to active duty in the aviation training program. First- and second-year collegians were enlisted in the V-1 program and might, at the end of their sophomore year, transfer to V-5 status. In April 1942, the Navy opened V-5 recruiting to high school seniors and graduates. By the end of the year, President Roosevelt had put an end to voluntary enlistment for men subject to the draft, and Congress had lowered the draft age to eighteen. The Navy promptly opened its doors to seventeen-year-olds and assigned young recruits to the new V-12 college program. As apprentice seamen they might be sent to college, where they "sported the bell-bottomed trousers and coat of navy blue" and had their way paid by the government until they could be transferred into the cadet program.[60]

The Navy's success in recruiting prospective aviators surpassed its ability to move them smoothly through a training program that had lengthened to at least eighteen months by the summer of 1944. Official estimates of pilot requirements had been revised sharply downward, thanks to a much lower combat casualty rate than had been anticipated and a failure to allow for the impact of combat fatigue; a reduction of the aviation training program seemed appropriate. The ax fell on July 5, 1944, in the form of a BuPers letter stating the Navy's determination to "effect separation from the flight training program of approximately fifty (50) per cent of all students now in Pre-Flight, CAA-WTS, Flight Preparatory, and Tarmac stages of training." As these cadets were in good standing with regard to academic and flight requirements, the blow was softened by suggesting that men with collegiate experience might be reassigned to the V-12 or V-7 line officer training programs. Others among the unfortunates might find a place in some less glamorous aspect of naval aviation. All were assured of readmission to the cadet program should the Navy need them in the future.[61] This drastic cutback coincided with the elimination of both the NFPS and WTS stages, the latter in August, the former in September, leaving Pre-flight the introductory phase of training.

The reverberations of the 1944 "deselection" announcement rolled menacingly through aviation cadet ranks. Some 7,000 cadets were soon affected by the separation order, and the disappointment of those who were shipped to the Great Lakes Naval Training Center for reclassification was scarcely alleviated by official explanations. Unfortunately, within six months the Navy discovered that its reduced estimate of pilot requirements had been in error. On January 1, 1945, the Bureau of Personnel announced that "recent studies of probable future requirements, made in the light of later experiences, indicated the necessity of increasing the scheduled future output" of pilots by 50 percent.[62] It was, in the vernacular of the day, a real snafu. Happily for all concerned, the war would be over in another eight months.

When Robert R. Rea enlisted in the Naval Reserve V-5 program in November 1942, virtually all the training phases developed for World War II were in place. A more detailed picture of each phase may be drawn by examining those schools and bases to which he was assigned.

That same November, William Jewell College, the site of Rea's NFPS training, opened negotiations with the Navy for a Flight Preparatory School unit. Acting President W. I. Hester had one eye on the prospect of a major decline in his modest student body of about 450 and the other upon a glittering opportunity. The Liberty, Missouri, college, with nearly a century of educational experience in the liberal arts, signed a contract in December, providing that the college would furnish meals, living quarters, and ground instruction for 600 resident cadets who would undergo a three-month-long program in entering monthly classes of 200 men. Lieutenant W. O. Bice, a veteran of eleven months in the Navy, was assigned to command the unit with an initial staff of three officers and two enlisted men. Eighteen William Jewell faculty members were engaged to instruct the cadets on a part-time basis; eventually forty-six new teachers were employed. President Hester declared, "We felt that this was our patriotic duty," and to his Board he observed, "It enables us to carry on our full program of Liberal Arts work." The Navy contract contributed the then handsome monthly stipend of about $37,000 to the college—and that was a bargain for the Navy.[63]

"Tramp, tramp, tramp—the first group of 200 naval aviation cadets came marching through the streets of Liberty last night from the Burlington depot to Brown gymnasium and then to New Ely Hall on the William Jewell College campus," reported the *Liberty Tribune* on January 7, 1943. As the tired cadets passed Colonial House, a girls' dormitory, they were serenaded by the coeds in tones that lifted their hearts. In the first of three dorms to be set aside for the NFPS battalions, New Ely, the young men spent the next several hours racing up and down stairs, hauling up mattresses stored in the basement for the Army cots which Lieutenant Bice had been forced to beg from the Olathe, Kansas, Naval Air Station, and making up their beds with two sheets and two white blankets. The next day the cadets would learn how to "two-block" a sheet the Navy way. Clothes went on coat hangers that had been hastily collected by the Cub Scouts of Liberty. When all was reasonably shipshape, the cadets mustered for sandwiches and cocoa in the dining hall.[64]

Liberty, Missouri, with a population between three and four thousand in 1943, was the county seat of Clay County and lay about fifteen miles northeast of Kansas City. It had a rural orientation—the surrounding farms took pride in their potato crops—but Liberty also boasted a cinema, civic clubs, a radio station, and other hallmarks of middle America. By the summer of 1943, a USO Center had been established in a former bank building on the town square and provided cadets with a dance floor, "coke bar," ping-pong tables, and a reading area. The officers at the NFPS were understandably anxious that their charges should behave like gentlemen; smoking was restricted to barracks on campus and forbidden on the streets of Liberty. Only soft drinks were served at dances. When the chaotic laundry situation led one cadet to vent his wrath in public, Lieutenant Bice warned the battalions, "If Cadets desire the continuation of their downtown privileges each and every one will see to it that he *act* the part of a gentleman at all times."[65]

Civilian students and Navy cadets pursued their studies harmoniously throughout the spring of 1943; in fact a shortage of uniforms made them almost indistinguishable until summer, when Navy civvies disappeared and cadets finally donned khaki shirts and pants, black shoes, and ties. The college maintained "the regular curriculum for its regular students," and President Hester assured his Board that "the Navy has not 'taken over' the college." "The two student bodies have in no way interfered with each other and . . . combined social events have delighted everyone." A cadet newspaper, the *Fledgling,* appeared, and in the fall of 1943, V-5 cadets provided replacements for the depleted ranks of the William Jewell football team.[66]

Naval Flight Preparatory School cadets were organized as a regiment of three battalions of 200 men each, with a cadet regimental commander and staff. Battalions were subdivided into platoons. Regimental formations were held every morning and evening, and platoons marched to and from barracks and classrooms to the sound of the "Navy cadence." Cadets brought in from the fleet were readily distinguished from the collegiate boots by their drill-field éclat and salty language. Their combat ribbons set them apart as supermen in the eyes of midwestern boys, who were occasionally given ten-day emergency leaves to help harvest wheat on their families' farms.[67] The platoon leader of 6B2, Rea's unit, was a magnificent drillmaster, an enlisted seaman whose "Marine cadence" could be heard from one end of the campus to the other and made his collegians tremble. He was a swashbuckler, though no swordsman; his run-in with a competent fencer (as close to a duel as Rea ever wished to fight) had the benevolent effect of demonstrating to the platoon that a "swab-jockey" was not necessarily a better man than a college kid.

Primarily a ground school, Flight Prep had a curriculum that was extensive, if not especially rigorous: simple mathematics and physics preceded an introduction to the mysteries of internal combustion engines, aerodynamics, and aerology. The communications course demanded increasing competence in receiving Morse code and reading wigwagging semaphore flags and blinking signal lights. Navigation and recognition were of greatest interest and immediate pertinence to a distant war. The Navy's wind blew from the opposite direction from civilian winds—with disastrous results in plotting-board ground speeds. Friendly and enemy aircraft, easily distinguishable from any angle when carefully studied, looked alike when seen in a fiftieth of a second or less. A grading scale of 4.0 points (perfect) set the failing mark at 2.5; some cadets took pride in approaching the former, others consolation in achieving the latter. Differences in academic standing might affect a cadet's weekend liberty or the date of his departure for the next stage of training, but they were frequently counterbalanced on the drill field or the obstacle course. In fact, a healthy spirit of cooperation among cadets quickly emerged—one for all, and all for one— whether the current "enemy" was the relationship of airfoil lift to drag or the sadistic zeal of an overbearing PT officer. Physical training was highly touted by the Navy but was taken lightly by most cadets. It sufficed that one could swim and do the requisite number of push-ups and pull-ups. Future pilots became adept at hand-simulated flight maneuvers, for the school's Link trainer was seldom operative.[68]

In September 1943, the new Tarmac program, and a reduction of the number of cadets drawn from civilian life, led to rumors that NFPS was on its way out. Admiral McCain squelched the scuttlebutt, and Flight Prep Schools continued to function for another year, but the last cadets left William Jewell on September 28, 1944. The officers and enlisted personnel departed in October.[69]

With both Flight Prep and War Training Schools distributed across the nation, it was common for cadets to remain in the same geographic area for both programs. The Fifth and Sixth Battalions at William Jewell were assigned to WTS units scattered throughout Kansas, Nebraska, and Iowa when they completed the course at Liberty, Missouri. The Navy, in its wisdom, also dispatched to Lawrence, Kansas, Californians who had never been east of Hollywood. The University of Kansas provided them with ground training for the next three months, and the Ong Aircraft Corporation taught them to fly at the Municipal Airport.

Lawrence was more cosmopolitan than Liberty and the University of Kansas vastly larger than William Jewell. In 1943, Lawrence was a town of over 14,000 inhabitants, nestled on the banks of the Kaw (or Kansas) River and looking up to Mount Oread (familiarly known as "the Hill"), the site of the state university since 1866. The historic antecedents of Lawrence were those of the pre–Civil War free-staters, the Jayhawkers. Lying forty miles west of Kansas City, Lawrence was fundamentally a college town, though a sprawling new powder plant scarred the fields across the river.

Because it already offered aeronautics courses, the University of Kansas was one of the twenty schools that had inaugurated the CPT program in 1939. When Pearl Harbor put an end to student demonstrations for longer holidays, the university began to attract a number of military programs: soldiers of the Army Specialized Training Program (ASTP), Navy machinist mates in bellbottoms (some 3,000 between 1942 and 1944), and V-1, V-7, V-12, and V-5 trainees in smaller numbers.[70]

The War Training Service cadets were quartered in empty fraternity houses—the TKE house for one—and enjoyed many of the amenities the brothers had left behind. To their khaki uniforms, the cadets now added green twill trousers, green windbreaker jackets, and green overseas caps, certain guarantee that they would not be mistaken for ordinary GIs. With only two officers attached to the unit and seldom visible, discipline was relaxed, though the Navy cadence rang defiantly across the campus as the cadets marched to class.

Ground school was a continuation of Flight Prep studies: navigation more advanced; aerology more practically oriented to the concerns of a pilot; Morse code more rapid (fourteen words per minute) and in baffling five-letter groups—presumably some secret Navy code that defied deciphering; more planes to be recognized in one seventy-fifth of a second; more engines to be studied as if the Navy had no mechanics; PT of course. More pertinent was the study of CAA regulations and principles of flight, for every morning and afternoon the bus left for the airport—weather permitting.

The Elementary phase of flight training in 65-horsepower J-3 Piper Cubs consisted of forty lessons: twenty-two hours of dual and check flights, fourteen hours of solo, in the case of Cadet Rea. Most cadets soloed after eight or nine hours of instruction. The Intermediate course at Lawrence introduced cadets to the Navy-built N3N, but flight time depended upon the availability of planes, instructors, decent weather, and luck. Bob Rea was able to squeeze in only six hours of dual and nine hours of solo flying in the 235-horsepower plane.

By October 1943, the Navy Department foresaw that sometime in the next year it could dispense with the WTS program, because the Navy would then be fully capable of conducting its own pilot training. On July 28, 1944, a program that had served the country well, in peace and in war, came to an end.[71]

From their initiation into the joys of flying, cadets were plunged into the rigors of Pre-flight Schools, of which that at the State University of Iowa was the oldest. The Navy had inquired about the university's interest in the new program in January 1942, and later that month a team led by Lieutenant Commander Tom Hamilton visited the campus to inspect its facilities. Early in February, President Virgil M. Hancher and other university officials went to Washington to discuss arrangements, and on February 6, Iowa became the first institution selected by the Navy for a Pre-flight School. Like other administrators, Hancher took pains to assure the public that the financial benefits that Navy Pre-flight would bring to the university would not threaten the integrity of its academic program. He stressed the physical separation of the two functions on campus and assured nervous constituents that the fine personal qualities of Naval Aviation Cadets were a sufficient safeguard against the womanizing and boozing propensities traditionally associated with sailors. The cadets were, after all, training to be officers and gentlemen.

Iowa City, the home of the state university, straddled the Iowa River and had been founded as the territorial capital in 1839, though it yielded that role to Des Moines after Iowa became a state.[72] Iowa was one of the country's better state universities and was a proud member of the Big Ten athletic conference. The contract signed by the Navy and the Iowa State Board of Education provided that the Pre-flight unit should have the use of two men's dormitories and partial use of the massive field house and adjacent drill fields. The university would supply food services for the Navy unit, and the government initially agreed to pay $1.21 for each cadet's daily food; this figure was adjusted upward as the price of commodities rose in the course of the war. A new classroom building would be erected at the Navy's expense, and the university would be reimbursed for remodeling the field house, 80 percent of whose space would be controlled by the Navy.

There was inevitably some friction in the generally satisfactory relationship between the Navy and the university. When the Navy tried to secure exclusive use of the field house, President Hancher balked, declaring that the university must continue to participate in Big Ten athletic competition. Hancher also clashed with the commanding officer of the Pre-flight unit, Captain David C. Hanrahan, an old salt who had commanded the Northern Bombing Group in World War I, over the availability of alcoholic beverages in the Ship's Store, which was located on university property. Hancher insisted that legislators and alumni were adamantly opposed to the sale of liquor, and he suggested that its availability could not be described as essential to the operations of the Pre-flight School. Hanrahan grumbled but eventually gave in. The Pre-flight School at Iowa was "the most dramatic and widely publicized of all the State University's wartime training activities" and was not disbanded until December 1945.[73]

In terms of its male students, the University of Iowa was virtually a military college from 1943 to 1945. The civilian student population sank to around 1,000—123 men and some 800 women. The Pre-flight unit counted 2,000 cadets on campus at all times. Women began to dominate many aspects of regular campus life. Coeds edited the student paper, the *Daily Iowan,* and the *Law Review.* Girls filled the ranks of the marching band. An all-male pep organization, the Scottish Highlanders, changed sex overnight. Women, for the first time, shared equally in student government. An alumna of those years recalled that Iowa became an essentially "female campus." For some of the girls who remained behind, there were tea dances and dates with cadets, but for more, the young men who had left the campus to go off to war were "boys from our home towns; they were part of the college we had hoped and planned for. When they left . . . part of our dreams for the future went with them; broken dreams that couldn't be mended by the khaki and navy blue."[74] Most of the men in navy blue were much too busy and closely restricted to quarters to enter into any part of campus life.

In one way, however, Navy Pre-flight contributed mightily to the scene. With primary emphasis upon physical training, with football all but the official technique for promoting teamwork among future pilots, and with a host of famous college coaches and players commissioned and assigned to Pre-flight Schools, they competed exceedingly well at both intercollegiate and interservice levels. On autumn Saturday afternoons, the familiar sounds rose from the football stadium: blaring bands, screaming cheerleaders, raucous crowds (although the swarm of automobiles was thinned by gas rationing). Appropriately named the

"Seahawks," the Iowa team, coached by Don Faurot, posted a 9–1 record in 1943. They lost to Notre Dame, 14–13, but they were victors over the Great Lakes Navy team which defeated the Fighting Irish. The Seahawks ranked second in the final national poll. In a photographic essay on Naval Aviation training, Pre-flight was portrayed as a varsity football team charging from the dressing room toward the camera, ready to meet the enemy.

Rear Admiral Louis E. Denfeld proclaimed that "the purpose of this School is . . . to provide military indoctrination, physical conditioning and essential ground training preliminary to the commencement of actual flight training at Naval Reserve aviation bases."[75] A host of brand new line ensigns saw to it that cadets stuck to that purpose. From prisonlike quarters in Hillcrest and the Quadrangle—spartan with Navy bunks, dressers, and desks—cadets mustered early and returned late from the playing fields and classrooms. Standard dress consisted of khakis in warm weather, green gabardine pants, khaki shirt, and an unlined blue windbreaker in the winter. A green version of the seaman's pea jacket also appeared. Dress blues and whites were issued and worn on liberty, and virtually every cadet seized the first opportunity to have his picture taken while wearing his new officer's hat with its anchor insignia.

Physical training was literally the first event of the day, and the last. Arduous calisthenics provided a warm-up for football, soccer, basketball, volleyball, gymnastics, boxing, wrestling, hand-to-hand combat, and swimming. Iowa's Olympic-sized pool could hold hundreds of cadets in its frigid waters, and competence was required. The day's PT was topped off by a session of some competitive sport, each cadet being assigned according to the whims of the athletic establishment. Any deficiency in athletic prowess was subject to correction: the "subsquad" for those who failed to swim "the Navy Way," and the "tree squad" (delinquents' names were once posted on a tree somewhere) for those who were otherwise deemed unfit. It would be fair to say that the fit survived and the unfit suffered. All apparently passed the final tests, even the agonizing, weight-carrying "step test" at the end.

The classroom was not neglected. The curriculum of NFPS and WTS was little changed except for the introduction of celestial navigation. The course was taught by highly qualified Navy instructors and presented a more rigorous challenge than any other part of ground school. Communication speeds increased, and ships received priority over aircraft in recognition training. Naval history and traditions were taught. The scriptures of *Naval Regulations* were opened, disclosing the peacetime secrets of protocol and calling cards. There was also target practice with a .22 rifle, but cadets drilled with dummy wooden rifles and plastic bayonets.[76]

Following the completion of Pre-flight, cadets were assigned to one of sixteen Primary bases scattered about the country. Historically the domain of the venerable "Annapolis of the Air," NAS Pensacola, Primary training rapidly expanded to encompass old Naval Reserve Air Bases, the huge Corpus Christi facility, and newly created Primary bases, all designated Naval Air Stations in 1943. At these installations the romance and the reality of naval aviation came together; civilian surroundings and instructors disappeared, and cadets faced the professional nature of their service for the first time.

Cadets like Bob Rea, who went from Iowa Pre-flight to NAS Olathe (Kansas) for Primary training, found a base with an old tradition and a new location. Established as a Naval Reserve Air Base at Kansas City, Kansas, in 1935, the training facility was moved in 1941 in order to escape the congestion both of civilian traffic and of the North American bomber plant at Kansas City. The Navy acquired 640 acres of prairie farmland some twenty miles southwest of Kansas City near the small town of Olathe. The site had once been a camping place on the Santa Fe and Oregon trails and more recently an auxiliary air field for the C.A.A., who had abandoned construction of a runway in 1941 because of excess moisture in the soil. Navy contractors overcame that difficulty, and when completed, NAS Olathe boasted a gleaming white pattern of two one-mile-long runways (eight inches of concrete over four inches of crushed rock) 200 feet wide, two 100-foot-wide taxiways, and a pair of parking aprons 1,260 feet long and 340 feet wide. (Cadets were, naturally, restricted to the gravel surface of the field for takeoffs and landings.)

Buildings rose rapidly—brick and steel hangars, flimsy wooden barracks and classrooms, stout-timbered "quonset-type" gyms and field houses. First there were cadet barracks, mess hall, and two ground school buildings, then a recreation building, drill hall, central stores and issue building, armory, dispensary, laundry, fire station, and public works building. The base had its own power plant, water filtration and sewage disposal systems, gasoline storage and pump house, and garage. The functions of the administration building spilled over into the brig. Five massive hangars dominated the flight line, overlooked by the control tower and flanked by the parachute loft. The station was designed to handle 1,000 students at one time, and at its busiest, the complement included 395 male officers, 2,193 men, and 228 Waves.[77]

When Av/Cad Rea arrived at Olathe in February 1944, the base was under Commander H. J. Brow, a twenty-seven-year veteran who, between the wars, had raced planes for the Navy, had set one speed record, and had commanded a cruiser scouting wing. Flying took precedence over all else at Olathe, although Kinnick Hall was big enough to contain five regulation-sized basketball courts, a fifty-meter swimming pool, dressing rooms, showers, and other facilities. The base also sported four full-sized football fields, four soccer fields, a baseball diamond, and six tennis courts.[78]

Ground school studies followed a well-established pattern. Celestial and plotting-board navigation problems grew more intricate, and aerology more pertinent, for Kansas weather was extremely changeable and decisively affected flight training progress. Communications classes were decorated with signal flags and taught the fine distinctions between Roger, Wilco, Over, and Out. Recognition was becoming an art, thanks to instructors trained at the specialist school established by the Navy at Ohio State University and capable of teaching rapid and accurate identification of friendly or enemy planes and ships.[79]

When flight operations began at Olathe, there were 117 training planes—NPs, N3Ns, and Stearman N2Ss—and an equal number of instructors. By mid-1944, the number of trainers, now largely N2S-5s, had risen to nearly 900, with over 300 instructors. The flight syllabus was divided into four stages. Stage A provided familiarization with the trainer, proceeded to dual instruction in such basic maneuvers as taking off and landing, stalls and spins, slips and S-turns to emergency landings, and ended with the solo flight, for a total of eleven hours. Stage B was divided between fourteen hours of dual instruction and nineteen hours of solo practice and check flights. Greater precision in slips and S-turns to emergency landings was expected at this stage. Slip landings were relatively easy, but S-turn approaches to the small landing circle were tricky. Other maneuvers in Stage B included the falling leaf, pylon 8's, and lazy 8's flown across the horizon line. In stalls and spins, precise control of heading was required. It was difficult for students to take spins seriously, but carelessness could lead to accidents.[80]

Cadets who passed the B check went on to Stage C, which was divided into thirteen hours of dual instruction and twenty of solo practice and check flights. Aerobatics dominated this phase and included snap roll, slow roll, Immelmann turn, split S, chandelle, and loop. As in

Stage B, emergency procedures received constant attention. D Stage, the last in Primary, included fourteen hours of dual instruction and nineteen solo hours and was devoted to formation and night flying. Students were taught the old-fashioned stepped-up echelon and vee formation, flown in World War I, wherein each wingman maintained station above and behind his flight leader and kept position during takeoffs, turns, and landings. Six hours of night flying were restricted to the main field traffic pattern and required no more than touch-and-go landings.

At first glance, the number of Primary training fatalities at Olathe seems quite high. Fifteen cadets and fourteen instructors died in accidents involving solo and dual flights. On two occasions, both cadet and instructor perished in the same accident. One additional Marine officer-trainee died in a crash. However, when the 463,220 syllabus hours flown by 4,500 Primary students at Olathe is taken into consideration, the fatality figures do not seem so appalling.[81]

The training plane commonly used at Olathe (and elsewhere by both Army and Navy) was one of the most famous of World War II—the Stearman Kaydet or Yellow Peril. From the perspective of forty years, former Av/Cad Rea remembers the N2S fondly.

It was a beautiful thing to look at, from its bold engine, down its lean flanks, the ribs showing, to its smooth tail assembly; its open cockpit was just where every kid sat when he dreamed of flying. The roar of the engine and the noise of the air all but destroyed communication from instructor to student, even when the Gosport speaking tube was used. The plane liked to move fast on the ground, and that led to numerous wingtip accidents and an occasional ground-loop. It was always something of a relief to get her either up in the air or dead in the chocks. But she more than made up for her feisty habits when airborne.

The N2S trainers we flew had wing spoilers so that they would stall and spin easily; they were lovely acrobatic planes. Sensitive to the controls, you could make them do almost anything, although neither my instructor nor I succeeded in doing an outside loop; we both ended up in inverted Immelmanns because the engine would sputter when you attempted inverted flight. One otherwise never felt the plane was to blame if a maneuver failed. They were great for precision landings by dead-stick slips but tended to float on S approaches (wherein the slip was forbidden). This occasionally led to proof that they were sturdy little aircraft. And normal Navy landings were great fun when a little power could be used; we were making one-point (tail-wheel) landings before we were through. The essential simplicity and responsiveness of the plane bred confidence. Regardless of the day's assignment, it was always fun to fly the Yellow Peril.[82]

World War II produced many changes in the training of naval flight personnel, including the separation of Primary and Intermediate training bases. While Primary training was dispersed among sixteen Naval Air Stations around the country, Intermediate training was concentrated at the Pensacola and Corpus Christi Naval Air Training Centers.

Before 1941, the history of Naval Aviation training could virtually be described as the development of flight instruction at Pensacola. The facilities of that glamorous "Annapolis of the Air" had been expanded through the emergency relief programs of the New Deal, and the Roosevelt administration's emphasis on air power had increased the number of aircraft on the base from 100 in 1935 to 400 in 1940. The economic and social impact of the Navy's presence was obvious in Pensacola before the war, and the Main Station, situated well away from the heart of the city, had become the hub of a growing town called Warrington, a community of old service and new Navy families. A network of auxiliary fields stretched out for forty miles east and west of the Red Cliffs on Pensacola Bay.

The auxiliary fields numbered six during the war. Barin Field lay to the west, across the Perdido River, in Alabama. Commemorating a crew member of the NC-1, "Bloody Barin" specialized in advanced fighter training. Bronson Field, on Perdido Bay, was named for Naval Aviator No. 15 and was the site of advanced divebomber training. Northeast of Perdido Bay, Saufley Field honored a veteran of the Veracruz expedition and was one of two auxiliary fields used for Basic training. A few miles to the northwest of Pensacola, Corry Field enshrined the name of Naval Aviator No. 23, the first native Floridian trained at Pensacola. It was the center of advanced torpedo training. On Escambia Bay, Ellyson Field proudly bore the name of Naval Aviator No. 1 and provided Basic training. Several miles to the north of Milton, Florida, was Whiting Field, named for a pioneer in carrier aviation and used for instrument flying and multiengine bomber training. At Main Station, Chevalier Field, named for the first pilot to land on the old USS *Langley*, was used for patrol boat training. Pensacola also boasted the Navy's only School of Aviation Medicine, where both flight surgeons and aviation medical examiners qualified for flight surgeons' wings. Headquarters for the Naval Air Training Command, established in December

1943, Pensacola provided the setting where almost half of the 28,000 Naval Aviators trained in World War II received their wings.[83]

An international cross-section of Allied trainees and instructors was highly visible on Pensacola's busy Palafox Avenue or on Main Station's freshly trimmed lawns, where moss-wreathed live oaks and permanent buildings gave the impression of an exclusive college campus. The white flashes on RAF blue forage caps and the white bands encircling dark blue Royal Navy Fleet Air Arm sailors' hats identified the largest foreign contingent. There were also English-speaking Free French students. The whole range of U.S. servicemen mingled with the foreigners, including Army trainees learning to fly P-boats for the AAF's air-sea rescue service and Coast Guard students taking similar training.[84]

In marked contrast to Pensacola, the Naval Air Training Center at Corpus Christi, Texas, was a war baby. While the ramps, quays, and fields at Pensacola had been developed over several decades, the Navy's installation at Corpus Christi rose almost overnight. Construction began in 1940, at the site of a tiny settlement on the shore of the bay, a dozen miles southeast of the city. "Residents were astonished as their little fishing village became a bustling town, and pelicans and rattlesnakes disappeared before the push of heavy machinery." Until a water main could be installed, Corpus Christi trucked water to the construction site, and the city improved dusty roads leading out to the station area. With Texas Congressman Lyndon B. Johnson (soon to become a Navy officer himself) in attendance, commissioning ceremonies were held on March 12, 1941. The Main Station, on Corpus Christi Bay, had wide paved streets, permanent buildings, and the whole sprawling complex of structures characteristic of a great military base. Pensacola might train more pilots than did Corpus Christi, but Corpus could boast the distinction of being "the largest naval aviation training activity in the world. Located in a state for which many superlatives are claimed, Corpus Christi sprawls over three counties, covers approximately 20,000 acres, and has 997 hangars, shops, barracks, warehouses, and other buildings."[85]

A major portion of the Corpus Christi training activity took place at six outlying auxiliary bases: Rodd was constructed first (1941), followed by Cabaniss (1941), Cuddihy (1941), Waldron (1943), and, near the town of Beeville, Chase (1943)—all named for deceased naval aviators. The sixth major auxiliary field, Kingsville, was completed in 1944 and named for the adjacent town; it lay on land that had once belonged to the famous King Ranch. The Naval Air Training Center also used a civilian airport at Rockport, across the bay from the Main Base, whose three concrete runways had been constructed by the C.A.A. Eighteen practice landing strips completed the system. Each part of the Corpus Christi facility served some specialized purpose. Basic training took place at Cuddihy and Cabaniss, instrument training at Chase, torpedo training at Rodd, advanced fighter training at Kingsville, divebombing at Waldron, and P-boat and observation instruction at Main Base (although training squadrons were occasionally transferred from one auxiliary field to another).

At Corpus Christi, as at Pensacola, students were organized as a cadet regiment composed of twelve battalions, and those who reached the Advanced stage of training were distinguished from their juniors by a double-bar cap insignia. Because of the constant movement of cadets between Main Base and the outlying fields and the inevitable periods of waiting in "pools" at Main Base, military exercises and discipline tended to be relaxed, in spite of the official view that "regulations prescribe strict procedure for nearly every detail of the aviation cadet's life." If ranking officers at Main Base sought to maintain a taut ship, at least one outlying field was unofficially known as "the Country Club."[86]

Physical fitness had, presumably, been established at Pre-flight, therefore it was only necessary that "fitness programs . . . continue in modified form, with other recreation type athletics thrown in to keep cadets . . . from going stale." There was one exception. It was officially held that, "of all the sports the cadet participates in at Corpus Christi, swimming ranks first in importance." Both speed and endurance were regularly tested—and encouraged by harrowing tales of the survival of downed pilots in the Pacific. Cadets experienced simulated ditching, escaped from cockpits under water, inflated Mae Wests and climbed into rubber rafts, and were impressed with the utility of shark repellents and marker dyes. Otherwise, PT was largely a matter of pick-up games played for fun as much as for fitness, the only requirement—that a ball be involved. It was a shock to cadets to discover that just prior to graduation they were once more required to pass the happily forgotten Pre-flight physical with its chin-ups, push-ups, obstacle course, and step test.[87]

Much of the ground training at Corpus Christi was sandwiched in between assignments to Basic, Instrument, and Advanced flight training bases. It followed the usual pattern but reflected the specialization toward which cadets were moving. No one believed that pilots would be expected to repair a constant-speed variable-pitch propeller, but its relationship to gas consumption was obviously pertinent. Radar was treated as a hush-hush subject, although it had been in use since the first day of the war. Cadets solved navigation problems more simply once they had the small Army pocket computer, even as their instructors complicated the problems by simulating realistic operational situations—from ready room to landing pattern. The importance of precise navigation was made clear by navigation flights over both land and water, simple patterns but sufficient to produce a rash of embarrassed, lost cadets. Gunnery introduced students to the reflector sight, first for trapshooting, then for .30-caliber machine guns. Synthetic devices such as the "Gunairstructor" simulated a fighter cockpit and a hostile sky. Silhouettes of enemy planes flashed across the screen, targets for electronic bullets. Student and instructor could battle it out and check the number of hits at the end of the duel—but it was difficult to overcome the Penny Arcade atmosphere.[88]

This training aid was but one of many that were perfected and distributed by the Navy's Special Devices Division. A shortage of operational equipment for training purposes early in the war, coupled with the tremendous expansion of Naval Aviation training, provided the impetus for developing new instructional equipment. The driving force behind the Special Devices Division was M.I.T.-trained Luís de Flores, a brilliant industrial inventor who returned to naval aviation for the duration of the war.

Perhaps the most valuable training device in the Division's arsenal was the Link Trainer, already more than a decade old when the war began. Available commercially in two models, the Link Instrument Trainer and the Link Celestial Navigation Trainer, the machines were used extensively to train domestic airline pilots but had largely been ignored by the armed services before pilots, airlines, and Links were virtually drafted into the service of their country. The Link Instrument Trainer was used in the Instrument Squadron at Chase Field, Beeville, to prepare cadets for blind flying "under the hood" and to teach them the basic skills of radio range procedure. The control panel of the Link Trainer was manned by an operator, often a Wave, and the student sat in a mock-up fuselage and cockpit under a lightproof hood. The trainee-pilot had before him the normal single-engine fighter controls (stick, rudder pedals, throttle) and the standard instruments of the type of aircraft he was being trained to fly (magnetic and gyro compasses, needle-and-ball and gyro turn-and-bank indicators, air speed indicator, altimeter, clock). He was also equipped with earphones and a microphone. The trainee "flew" the Link in response to oral instructions received from the operator and in accordance with the radio range code signals he heard. He could simulate all normal flight maneuvers from takeoff through climb, turn, and descent and could do so at any designated rate or speed. Radio range practice included orientation, holding patterns, and let-down techniques. Maneuvers were performed with various and changing wind conditions that caused the "plane" to drift off a given course and required correction of the compass heading. No radical maneuvers were practiced in the Link Trainer, but it would stall and spin if the pilot failed to maintain proper air speed. Spin recovery was possible, but novices were frequently chagrined to hear the operator chortling, "That's all—you just spun in!"[89]

Intermediate Flight Training was divided into two components, Basic and Advanced, which were further broken down into specialized segments according to types of training, availability of equipment, and the Navy's current pilot requirements.

Basic training introduced cadets to a modern, all-metal, low-wing aircraft, usually the SNV or Vultee Valiant (familiarly known as the Vibrator), with a 450-horsepower Pratt & Whitney Wasp engine. Retrospectively, Bob Rea observed:

> This was my first low-winged plane, a type approached with an excitement that quickly died away. The V was not a pretty plane, though obviously a stout one. For its size and weight, it was underpowered and therefore never performed up to one's anticipation, whether going up or coming down. It was not very maneuverable, was sluggish on the controls, but very stable—which was undoubtedly a good thing at this stage of flying. I never loved it nor knew anyone who did.[90]

In this plane cadets received dual familiarization instruction, soloed, and were taught formation flying in the new stepped-down echelon and vee, both in daytime and at night. Violent maneuvers were forbidden in the SNV, but extremely close formation flying, with positions constantly shifting as planes crossed under each other, placed significant demands upon the best of student pilots, particularly in night flying. The course was typically completed with ten hours of dual instruction and twenty-two hours solo.

At this point in his career, a cadet was allowed to indicate a preference for future specialization in multiengine bombers or flying boats, single-engine observation aircraft or carrier-based planes. Decision rested with the Navy, of course, which was apt to assign 70 percent of a class arbitrarily to training designed for future carrier pilots.

From Basic training, cadets were transferred to an Instrument Squadron where they were first given instruction in a Link Trainer and then flew (always dual) under the hood, perhaps fifteen flights in all for a period of twenty-one or twenty-two hours. If night formation flying had been nerve-racking, blind flying (to the accompaniment of an instructor's heavy breathing and sometimes frantic directions) was frequently excruciating. A cool head and a light hand were required; a compliment on a well-executed maneuver was more thrilling than the first solo loop, and the final up-check was perhaps the most gratifying moment in the whole training program. With the completion of Instrument Squadron, cadets shed their single-bar insignia and entered Advanced Training.[91]

In the final stage of flight instruction, cadets were trained in planes typical of their future service specializations. Those destined for multiengine land-based aircraft flew the Beech SNB trainer. Flying boat pilots went into the faithful old PBYs. Scouting and observation trainees were put into OS2Us. Cadets headed for carrier assignments flew the most famous of all World War II trainers, the North American SNJ, a plane with operational characteristics, competence, and history—a plane with everything, including guns. As Rea remembered it,

the J seduced me from my first love, the Stearman, and understandably; but she was more of a woman in every way, including fickleness. She looked the way a military plane should in the early 1940s. She had power, more than one ever really used, and she wanted to be flown with power—mastered—at all times, yet she demanded finesse. Her more sophisticated controls (propeller, engine, flaps) made landing and taking off interesting, not necessarily easy. A fast takeoff left little time for fine adjustments, with the result that confusion in formation takeoffs was fairly common. Once in the air, she scooted. Landing was always a bit exciting, for the syllabus required gliding approaches to a field, and the J was not a good gliding aircraft. When, after training, I could fly my own way (i.e. making a carrier approach to landing) landing the J was a delight (even if it scared my observer). It was easy to over-control the J, and that was undoubtedly the cause of most problems that occurred to all pilots. Much as we all liked the plane, she tended to keep you respectful.

In the air, the J was sweet, but aerobatics in a low-wing monoplane are different from those in a biplane, and available speeds led to a certain caution in maneuvering. The plane was perfect for what we were supposed to do, however, and performed beautifully in gunnery runs. The J's responsiveness also made for good instrument flying, where a light hand and foot got immediate response when necessary. All in all, the J seems to have been an ideal training plane in every respect, making the transition to F4U Corsairs as easy as possible, for virtually all the characteristics of the heavier plane had already been experienced—and additional power never hurts.[92]

The flight syllabus in Advanced was strenuous and varied and included seventy-six flights totaling some fourteen hours of dual instruction and eighty-six hours of solo flying. Aerobatics practiced with combat application in mind rapidly produced an appreciation of the relationships between weight, speed, power, and stalling in violent maneuvers. The traditional vee disappeared from formation flying, giving way to the two- and four-plane section and division found best suited to combat conditions. Spatial separation, rather than perilous proximity, became the rule as radical maneuvers replaced stately processions through the sky.

Air-to-air gunnery entered the picture at last. After practicing dry runs under the eye of an instructor, cadets fired live ammunition at tow-sleeve targets, long cloth tubes attached by a trailing wire to a towing plane. The bullets of each plane were color-coded, red, blue, green, yellow, in order that each man's hits might be identified, and counting hits was a sure way to humble anyone who thought himself a marksman. Because tow wires broke under the stress of excessive speed, cadets always prayed for a patient, long-suffering tow pilot who was in no hurry to return to base and did not lose his target over the Gulf or drop it somewhere in the boondocks. Strafing practice was minimal, presumably on the assumption that any competent pilot could follow his own tracer bullets to the target.

Divebombing was the most difficult phase of Advanced, to the surprise of most cadets. The disastrous effects of slight variations in speed and dive angle, and most decisively in wind direction and velocity, were learned the hard way. The SNJs carried small dummy bombs under their wings, which were aimed at a 100-foot target circle which seemed to shrink rapidly in middive. Instructors were often inclined to credit near-misses (as naval gunners have always done), but an inflexible check pilot could cause consternation among the most eager students.

Their plotting-board navigation having been brought to the peak of practicality in ground school, cadets in Advanced training had the opportunity of testing their skills in practice flights over both land and water. Flown in four-plane divisions in which the lead and navigational responsibility changed with each leg of the flight, Nav hops presented simple plotting problems but useful tests of a cadet's ability to fly a straight course at a constant speed. Mechanics frequently flew in the rear seat of the SNJ in order to log flight time requisite to their flight pay; their presence relieved the tedium of flying a wing position on a predetermined heading and sometimes offered an opportunity for much wild chaffing over the intercom regarding the real and imagined hazards of flying, for young seamen were easily impressed by old jaded cadets. If flying from one little Texas town to another was usually boring, some cadets erred, discovered the Rio Grande River, even made forced landings at Army Air Force bases. Most cadets became more serious about navigation when they moved out over the Gulf of Mexico and beyond sight of land. Happily, as a station publication advised "pilots strange to this vicinity," "Corpus Christi is at a critical bend in the Texas coastline and . . . the Gulf of Mexico is generally east of South Texas."[93] Such truisms were worth noting.

A very few hours in Advanced were dedicated to combat tactics (individual plane against plane), section tactics, and flak evasion, but these activities involved little more than glorified tail-chases that whetted appetites for Hellcats or Corsairs.

By the latter part of 1944, a further stage of training had been added to the Advanced cadet's program. Known as Pre-operational, it familiarized prospective carrier pilots with that noble old veteran, the Douglas Dauntless SBD divebomber. Built in countless number, the SBD earned its fame in bloody combat and was retired in favor of the new Helldiver. The rather surprising result was a surplus of planes and the opportunity of giving cadets a taste of a combat aircraft. The Pre-operational stage was too brief and the old SBDs at Corpus Christi too much in need of repair to accomplish anything significant in the way of instruction, but five short flights (seven and a half hours) in a warplane was a thrilling and a sobering experience for cadets impatient to be commissioned and eager to enjoy the accompanying leave.[94]

Recruiting and publicity brochures depicting flight training at Pensacola and Corpus Christi invariably featured the culminating moment of cadet life—a graduating class parading in dress whites before a reviewing stand, the commandant presenting wings of gold to a newly commissioned ensign. No doubt there were such grand occasions when the heavens smiled and time stood still, but the young pilot's immediate future in the Navy had already been determined and orders had already been cut dispatching him to one of several Operational bases where training would proceed—albeit in rather happier circumstances. He would be an officer earning flight pay, and as likely as not he would also be a very recent bridegroom, for the Navy's prohibition of marriage for cadets ended with the commissioning ceremony.

Before 1941, Operational training had been conducted in squadrons attached to the fleet, but Pearl Harbor forced the Navy's postgraduate program ashore, and the first of the new wartime functional training divisions was the Operational Training Command, with headquarters at Jacksonville, Florida. Utilizing established bases at Miami, Key West, and Banana River, the Command soon embraced a dozen facilities adjacent to the east coast of Florida and southern Georgia. Squadrons stationed at each of these bases provided training in a particular type of warplane—fighter, divebomber, torpedo bomber, and so on; their common objective was to put the new Naval Aviator into a combat plane and teach him to use it as a weapon.

Because of combat requirements, the Operational Training Command initially faced severe material shortages; the latest versions of aircraft naturally went to the fleet, and ordnance was in short supply. Until wartime production overcame those problems, the Command met the challenge by extensive and innovative use of synthetic training aids such as the Gunairstructor. The assignment of experienced fleet pilots to these training squadrons introduced a further sense of realism and flexibility to a program which was the capstone of naval aviation training.[95]

The Operational Training Command's bases included Lee Field, Green Cove Springs, Florida, a civilian airfield taken over by the Navy in 1940 and used for a variety of purposes until 1943, when squadrons VF-1 and VF-3 arrived to direct the training of fighter pilots. The Naval Auxiliary Air Facility at Green Cove Springs was unpretentious but comfortable. The field was surrounded by swamps, but its four 5,000-foot asphalt-covered runways were ideally suited for Hellcats and Corsairs.[96]

The business of Operational training was combat preparation. After familiarization with the heavier, more powerful aircraft being flown, pilots began a rigorous course of tactical practice in section and division formations and in individual combat, gunnery, and bombing techniques. Overhead firing runs and high speed pull-outs introduced pilots to G-forces and blackouts. Camera guns disclosed weaknesses in combat tactics and proved that mere proximity to an enemy was not enough to achieve a kill. More thought-provoking was the observation that, out of thousands of rounds fired, a mere 2½ percent of hits on a target was considered adequate to do the job. Operational training was rigorous, but it was also rewarding. Veterans of the toughest aerial fighting of the Pacific war taught the meaning and value of leadership in the air, and experience demonstrated that the requirements of teamwork in flight were different from those on a football field. There were still unexplained service mysteries, however. In 1945, F4U Corsair pilots, flying the Navy's hottest fighter, were being trained in the technique of glide bombing, an attack flown at a shallow angle of approach and at a slow speed which could only be maintained by lowering both wheels and dive flaps. No one knew why; no one liked it.

Honing skills and polishing tactics was routine training; learning to land and take off from a carrier was a new experience. Before the war and the dramatic expansion of naval aviation, pilots made their first deck landings on fleet carriers operating out of East and West Coast ports. When such procedures were no longer practical, training carriers such as *Charger* at Norfolk or *Wolverine* and *Sable* on Lake Michigan were used. The war introduced a new class of small "escort" carriers (CVE), and by 1945, they and the old *Ranger* were available for training duty in coastal waters. Pilots in Jacksonville Operational Command made their first landings aboard the CVE *Guadalcanal*, a tubby little vessel with the proud record, under the command of Captain Dan V. Gallery, of having captured the German submarine U-505 in the Atlantic in June 1944.

Carrier qualification (CQ) began on land in an SNJ with a demonstration of the basic principles of flying low and slow. Pilots then had to practice flying their planes at just above stalling speed and landing on a runway in accordance with the directions of an LSO's (landing signal officer's) hand-held paddles. This Field Carrier Landing Practice also provided the experience of a catapult take off. When pilots met the FCLP standard, they were sent aboard "Fuzzy," the *Guadalcanal*, to make their daytime qualifying landings. That experience was memorable, for few boot ensigns had ever been aboard a U.S. naval vessel before, much less aboard a carrier. There was the time-honored protocol of salutes upon first entering the ship, the cold welcome of a ship's bunk in a tiny two-man cubicle, and the first nervous pacing of a flight deck as the 'Canal nosed into the Atlantic. Anticipation of his first carrier takeoff and landing was unquestionably the most serious moment in the training career of any Naval Aviator.

The event itself was not disappointing, although the activity on the flight deck of a carrier that was launching planes was a major distraction. Suddenly the plane was positioned for takeoff; the launch officer's arm was coming down. Throttle forward, brakes off, a shuddering, roaring leap forward—airborne. Once off, there was no visible option but landing on that tiny rectangle below—slowing, checking that the tail-hook was down, keeping the officially ordained speed in the turn, and straightening out to realize that the LSO's elevated paddles were virtually screaming, "Too high—too fast!" The first wave-off was at least as much of a relief as it was an embarrassment. The second approach was right, all the way to the "Cut!" The Corsair landed with an alarming thump but caught an early wire, and before self-congratulation could well begin, the plane was released and was being waved forward to take off again. The experience was one to strengthen a pilot's faith in God, the Navy, and himself.

Carrier qualification was the final phase of Operational training for fighter, dive, and torpedo bomber pilots. From there they might be assigned to an Air Group, most likely on the West Coast, preparatory to joining the fleet, or they might be ordered to specialized training in night fighters, the latest refinement in naval aviation.[97]

Ensign Rea's experience was typical of many during the last months of the war. He was attached to Air Group 98 at Los Alamitos, California, but arrived there only after being transferred from one auxiliary base to another. Los Alamitos NAS had been established in 1928 at the Long Beach municipal airport. In 1942 it was relocated, some six miles away, because of congestion at the Long Beach field. The new station reflected the usual architectural inelegance of wartime construction but provided more than the usual amenities. The foyer of the officers' mess hall even offered a wide selection of pinball machines and one-armed bandits ready to accommodate coins of every denomination. Incoming pilots sensed that they were transients in a permanent Navy establishment.[98]

Los Alamitos had been an E base, a Primary base, and was now the seat of a Carrier Air Support Unit (CASU), which meant that it collected, trained, and dispatched replacement pilots to the fleet. It provided minimal exposure to such things as airborne radar and airborne rocketry, neither of which seemed to have been brought to a very high degree of efficiency. More impressive were those occasions when AG-

98 flew practice missions with TBF torpedo bombers below SB2C dive-bombers; F4U Corsairs hovering above them in close formation, struggling to throttle back to the slow rate of climb of the bombers; Hellcats cheerfully scampering overhead in the role of fighter cover. These "group gropes" ended in a mock attack on the home field or, more realistically, in bombing and strafing a rocky offshore island in the Pacific. The former exercise clearly demonstrated the importance of planning and organizational exactitude, for the air could get dangerously full of planes when a hundred or so tried to attack the same spot simultaneously; the latter may have disturbed the wild goats who populated the target islands—environmentalists have subsequently secured the removal of that endangered species from their noisy habitat.

The most significant training provided by AG-98 was in night carrier landings, first in touch-and-go practice sessions on a small desert field safely removed from civilization, then aboard the carrier *Ranger,* operating out of San Diego. Landing technique was the same, day and night, but the loss of visual contact at night, for both pilot and LSO, introduced a factor of insecurity and increased the accident ratio dramatically. As the experiences of the pilots whom Ensign Rea accompanied aboard *Ranger* demonstrated, night carrier qualification was far from automatic—or even certain.[99]

At this point, pilots were assumed to be prepared to join the fleet, and as replacements for combat units, they were being ordered out to Pearl Harbor when the war in the Pacific came to its abrupt conclusion. It was obvious to all that there were yet lessons to be learned regarding the application of flying skills to combat conditions—lessons which could only be taught by men who were on the firing line. The Navy's training program had fulfilled its task, and the performance of the generations of cadets who won their wings at Pensacola and Corpus Christi was solid proof of its effectiveness. The experience turned boys into men, and men into pilots; it also taught them that there is always something more to be learned before mastery may be claimed—the lesson implicit in the best sense of that oft-repeated axiom: "There are old pilots and there are bold pilots, but there are no old, bold pilots."

# 1

## Personal Background

THE AUTHOR OF the following letters, Robert Right Rea, was born in Wichita, Kansas, on October 2, 1922, the only child of George Edgar and Fleda Schollenberger Rea. The city in which he grew up was young enough to remember its cowtown antecedents and brash enough to boast that it was "the Aircapital of the World" and to claim a population of 100,000. Robert's maternal grandfather had built a sod house on the Kansas prairie, and as a lad Bob met an elderly lady who had been kidnapped by Indians from her Wichita home and an old gentleman who had witnessed the bloody aftermath of an Indian massacre a few miles south of town. At the same time, Wichitans gathered by the thousands to watch the planes at the Travel Air, Stearman, Swallow, and Cessna factories, and they picnicked at the half-completed Municipal Airport while Ford Trimotors landed on a summer's evening. Wichita stood between its cattleman's past and a great aeronautical future—just as that generation stood between two world wars.

Like most Wichitans, the Rea family came from someplace else. The Reas were Hoosiers who had farmed above Madison, Indiana, on the Ohio River, since the late eighteenth century. George Rea (1881–1948) left Madison, and after trying his hand as a farm machinery salesman from Texas to the Dakotas, he settled in Wichita. The discovery that oil and water do not mix led him into a new business venture when petroleum-rich El Dorado was forced to turn to Wichita for bottled drinking water. Horse-drawn wagons could not compete with solid-tired trucks, and having successfully demonstrated the efficacy of the new mode of transportation, George Rea employed his equipment locally, forming the Rea & Bell Transfer Company, later the Merchants Van & Storage Company. As small businesses go, it was a success.

At about this time, George met Fleda High Schollenberger, organist at the Central Christian Church, where the bachelor businessman sang in the choir. The Schollenbergers had moved west from Pennsylvania to Albuquerque, New Mexico, where Fleda was born in 1882. The family subsequently settled in Wichita, running a small hotel which, according to a venerable traveling salesman, was the only decent place to eat in Wichita early in this century. Jacob D. Schollenberger, a patriarchal old gentleman, devoted his last years to building neighborhood churches (no matter what the denomination) and, as a staunch prohibitionist, bailing the hatchet-wielding Carry Nation out of the Wichita jail. Fleda was a professional musician and cared for her elderly parents until, in 1917, she married George Rea. Marriage during wartime would be repeated in the next generation, but George Rea was old enough to escape service in the First World War, and his family was well established in a comfortably roomy house when his son was born.

The household in which Robert grew up was run by his mother but dominated by his father. George Rea was a powerful six-footer, bald at an early age, who never raised his voice nor needed to, at home or at work, where his men respected the boss who could do—and did— anything he asked them to do. He fondly remembered his youthful prowess as a boxer. In his mature years he found relaxation in hunting and fishing, boasting that he never shot a squirrel out of a tree, save through the head, and never caught more fish than his family could eat. He was understandably disappointed that his son inherited neither his stature and physique nor his love of roughing it on the banks of the Ninescah River. Fleda Rea was a proper Pennsylvania Dutch housewife. She was house-proud, and her kitchen was the source of endless delight to her husband and son. Dinnertime was family time, and there an attentive lad might learn much. After dinner she introduced her son to the piano sonatas of Beethoven and accompanied her husband as he sang the love songs of their generation. Life was quiet, ordered, and disciplined, filled with a love carefully restrained by what were then considered to be the proprieties.

Robert Rea was educated in the excellent public schools of Wichita. As elementary and intermediate schools were no more than a block from home, his was very much a neighborhood society; friends and playmates lived within the elm-shaded confines of South Martinson Avenue. He moved rapidly through the grades and went to East High, on the far side of town, because it was relatively close to his father's place of business. Along with the evident advantages of good schooling, Bob enjoyed the hidden advantages of being put to the broom, the bucket, and then the loading dock of his father's warehouse. The boss would not allow, nor did the son receive, any privileges of status—that was to be earned if it was to be enjoyed, and truck drivers are great social democrats. At the same time, Bob was led into music by his mother and rather casually practiced the violin until, having achieved the principal's chair in the high school orchestra, he and his teacher agreed that he was not destined for the career of a professional musician. That decision opened the door to musical pleasure, and his collegiate years were filled with fiddling, singing, and long evenings devoted to absorbing the recorded classics in the company of friends. In high school, a boy's delight in the novels of Alexandre Dumas found active release in a self-taught fencing club. A relative latecomer to the Musketeers, Bob happily inherited the mantle of D'Artagnan and devoted himself to proving his merit with a foil. After watching his first intercollegiate fencing match, George Rea pardoned his son's disinterest in boxing, baseball, and other "manly" sports.

In the fall of 1939, Bob Rea began his studies at Friends University, a Quaker college in Wichita which had a fine reputation for scholarship, music, and broadly tolerant Christian values. Classes began on September 2; Hitler had invaded Poland the previous day. While adults who remembered the war to make the world safe for democracy seemed to disbelieve that such a war could happen again and were surprised and horrified by its outbreak, the entering college class of 1939 was not. For years they had read of war in Manchuria and other parts of China, Ethiopia, and Spain and had watched the swastika creep like a shadow across central Europe. Under the evening streetlights they had weighed the future, as boys will do among themselves, and concluded that war was inevitable. They would simply have to live with it, quite certain where right lay and equally sure that, as Americans, they would be there.

Happy the young man who can spend his college years in a small school wherein ideas flourish and intellectual competition is the norm, where the arts are nourished and individual initiative replaces team effort. That was Friends University between 1939 and 1943. There were social fraternities and sororities, such as Alpha Kappa Tau and Delta

Rho Alpha Nu, where town boys and dorm girls met and fell in love. There were sports, for Bob tennis and fencing and—surprisingly—an athletic letter. There was no drinking, no smoking, and the dances were officially unofficial. There was also a student newspaper, the *University Life,* on whose staff Bob began to write, turned to soliciting advertisements, and wound up as an editor—a position in which he learned the dangers of a too-free press. If he made *Who's Who in American Colleges and Universities,* he did not (quite) make the university honor society.

From the outset, the war posed a problem of values and understanding in a Quaker school with firm pacifist beliefs. Like the majority of his friends and fellows, Bob was neither Quaker nor pacifist and was inclined to scorn their views. As young men began to leave school for the armed services, the arguments became heated at times, but the point was made, on both sides, that good men may differ in matters of principle, and draftee, volunteer, and C.O. learned to accept one another.

The war remained distant until December 7, 1941. Bob and his friends heard the first radio reports of Pearl Harbor while hashing over the results of the previous night's fencing match. Then, for most college men, it became a question of how long it would be before the draft caught up with them. The armed forces instituted programs such as ASTP, V-1, and V-5, which allowed a volunteer to attend college while in service or to remain in school until graduation. It was clearly the way to go if one might choose. With the draft board breathing down his neck, Bob sought admission to the V-5 Naval Aviation program in October 1942. Why the Navy? Cleaner aboard ship than in a trench. Why aviation? Young Wichitans had grown up in "the Aircapital" (even if they had never gotten off the ground). And one of Bob's closest friends, William E. Roy, had just won his Navy wings of gold. Bill's whites outshone another friend's olive drab; his sword (even aviation officers had to have them in those days) felt good to a fencer's hand; and Bill knew the trick of polishing his shoes with gin. Enough to take one down to the recruiting station at the Post Office, where you learned to stretch past five feet six inches (Navy minimum, Rea's maximum) and to worry about the effects of too much reading on 20/20 vision. Massive ingestion of carrots took care of the latter concern (or was believed to help), and on November 4, 1942, Aviation Cadet Rea was sworn into the V-5 program, U.S. Naval Reserve, in Kansas City, Missouri. The previous evening he had written to his girlfriend:

Dear Phyl,

Pardon my failure to write Saturday as I usually do, but everything was hanging fire then and I was a bit up in the air. Of course I still am, but the fog is clearing somewhat. "Fog" in case you are curious, is my status with the U.S. Army. At present I am in the process of affiliating myself with the U.S. Naval Reserves as an Aviation Cadet. That will allow me to finish this year at Friends [University] and then give me 8 months training, possibility of a year as an instructor, and then—who knows?

I decided all this Monday [October 26] after receiving notice of my draft physical. I got my papers in order and am here a week later taking lots of tests and wasting two days. I had no trouble in Wichita except getting my eyes O.K.ed. I had read 10 hours Sunday, and Monday my right eye was about 75% efficient. It was alright Tuesday, however, and was fine this afternoon when tested. These navy recruiters are an interesting lot, some of the worst sour-pusses and gripers I ever met, others swell fellows. J—— and P—— S—— are up here now getting into the V-1 program. Their tests were over today and mine will be tomorrow. No hitch in the get along yet, but my fingers are crossed until I get all sworn in.

The Navy kept its word and allowed Bob to graduate from Friends University on May 31, 1943, with an A.B. in History. His Navy orders to report for training were already in hand. From this point his letters may speak for him.

The correspondence presented here consists of most of the letters written by Robert Rea to his parents between June 1943 and November 1945, when he was discharged from active duty and returned home. Like most wartime letters, they were intended to reassure as much as to inform. Consequently, they were couched in homely terms and dealt with matters the writer might assume would be of interest at home. They omitted those inevitable personal foibles that most young men prefer to hide. Otherwise they were open and candid and provide a reasonably complete picture of the conditions and nature of the naval aviation training program. Each letter was devotedly filed by the author's mother, secure in its envelope bearing the serviceman's postmark "Free," and remained untouched from 1945 to 1983, passing from her to the writer upon her death in 1958.

To Bob's letters to his "Dear Folks" have been added excerpts from his letters to Phyllis Jeanne Edwards, who became his wife on February 14, 1945. Phyl was the daughter of Dr. David Morton Edwards, a distinguished Quaker educator, formerly President of Earlham College, and

President of Friends University at the time of his death in 1939. She was then a sophomore, a leading figure in all aspects of campus life at Friends University. Brought together by common classes, interests, and activities, Phyl and Bob became romantically involved in 1941, only to be parted in 1942, when she left for graduate study at Boston University. At that point Bob began a correspondence which continued until May 1945, when Mrs. Phyllis Rea joined her husband at Green Cove Springs, Florida. Phyl and Bob remained together during the final stages of his naval training—and recently celebrated their fortieth wedding anniversary. It hardly need be said that their correspondence was not primarily concerned with naval affairs, but as far as it was, it fills some gaps in the first set of letters, and it portrays events in a somewhat different light, often plainer and harsher regarding matters more immediate to the generation that fought the war than to their parents. Curious readers may fill the scholarly omissions with the content of their own love letters.

# 2

# The Distant Drums of War

**W**HEN TWENTY-YEAR-OLD Robert Rea was sworn into the U.S. Navy's V-5 program on November 4, 1942, World War II had been in progress for three years, two months, and four days. In the year of his enlistment, the armed services put together an indoctrination film series entitled *Why We Fight,* based on old and recent newsreels. With riveting narration and drum-punctuated musical score, it unfolded the epochal events that had led to September 1, 1939, and those that had since reshaped the world. The film pictured Adolf Hitler working himself into hysteria before a crowd that responded with mechanical frenzy; a diving Stuka with keening siren and whistling bomb; a dead child in a blasted street; panzers on the point of the blitzkrieg in open country beyond the Meuse; the sky over Kent whorled with contrails as Sir Hugh Dowding parceled out Britain's "few" to intercept Goering's black-cross bombers. Through his periscope a smiling "Grey Wolf" captain savored the fiery results of his handiwork on an oil tanker. A begoggled Field Marshal Rommel paused in his command car to muse over the smoking hulks of British armor near Tobruk. On and on the images flickered in black and grey and white, reminding Americans of Pearl Harbor on that special Sunday in 1941 when a tranquil scene was transformed into a latter-day Dante's Inferno. No less grim were the subsequent losses of the carriers *Lexington* and *Yorktown* and the sky battles of the Coral Sea and Midway.

Midway occurred five months to the day before Bob Rea took his oath. From that short perspective it was not possible to realize that the "five minutes of Midway"—in which three Japanese carriers were fatally struck by U.S. carrier-borne Dauntless divebombers, to be joined later by a fourth carrier of the Imperial task force—"spelled the ultimate doom of Japan."[1] Defeat in the Coral Sea and at Midway turned the enemy to an overland offensive through the jungle and across the steep Owen Stanley mountain range of New Guinea. Their objective was the key base of Port Moresby, from which they planned to launch an invasion of Australia.

As a companion piece to their New Guinea campaign, the Japanese had begun to occupy the adjacent Solomon Islands. From August 1942 onward, the name "Guadalcanal" increasingly made banner headlines. It was the first major objective of an Allied counteroffensive in the Pacific, and it came to symbolize the classic rigors of jungle warfare as isolated U.S. Marines fought off attacks from land, sea, and air. The action focused on an airfield called Henderson. From there, Marine, Navy, and a few AAF pilots engaged the Zero, an enemy fighter superior in several respects to the U.S. Navy's and Marine Corps's Wildcat and the AAF's Lightning, the United States' first-line fighters of that day. In furious air battles the Japanese "suffered a significant weakening of air strength through the loss of trained pilots. . . . The Japanese could ill-afford this attrition, for American plane production and pilot training were increasing."[2] The climax of the struggle for Guadalcanal came late in October and during the first two weeks of November. When the final banzai charge was repulsed and the last Japanese effort to reinforce "the Island of Death" ended in costly failure, the Marine commander signaled to the Navy, "The men of [Guadalcanal] lift their battered helmets in deepest admiration."[3]

The Japanese were still entrenched on Wake, in the Gilberts, the Carolines, the Marshalls, and the Marianas. They held the Dutch East Indies, the Philippines, and Formosa. From Burma they threatened India, and squabbling between Washington and Chiang Kai-shek over American neglect of the CBI (the China-Burma-India theater) threatened to deliver China to the Japanese invader. Yet November 1942 was a time for relief, as in both the Solomons and New Guinea, Allied forces began to shove the enemy back.

The wheel of fortune also turned toward the Allies in North Africa. On October 23, the irascible British military hero Bernard Montgomery opened the battle of El Alamein with a thunderous artillery barrage that lit up the desert night. Montgomery "combined the stern talents of

Cromwell with the persistence of Wellington."[4] Having lifted the morale of the disheveled Eighth Army, he now gave it lasting fame in a smashing victory over the Afrika Korps of Field Marshal Erwin Rommel and terminated the threat to the Suez Canal.

On November 8, the Americans and British lit the TORCH of a three-pronged invasion of North Africa designed to effect a pincers movement with the advancing Eighth Army. It would not be a short campaign, an easy victory. Conflicts between the Vichy French, the Free French, the Americans, and the British added to battlefield difficulties. The final defeat of the Axis in North Africa would take almost six months, during which time the flamboyant and controversial George Patton would emerge as an American hero.

In the night skies over *Festung Europa,* RAF Bomber Command's streams of Stirlings, Halifaxes, and those greatest of British heavy bombers, the Lancasters, had mounted an increasingly destructive campaign of strategic bombing since early 1942. Yellow, green, and red flares, called "Christmas Tree," marked targets in Cologne, Hamburg, Bremen, and now and then, "just to show the Germans it could be done," Berlin. Once promised by their leaders immunity from this rain of death and devastation, the German population found that their will was put to the test. For RAF crews the cost was high as the sky was stabbed by searchlight beams, rent by flak, and made the hunting ground of Luftwaffe night fighters.

From English bases the AAF's Eighth Air Force had commenced a limited daylight, high-altitude, precision bombardment campaign of its own in the summer of 1942. British and even some American critics were skeptical of daylight bombing, and the former claimed that the much publicized Boeing B-17 Flying Fortress was inadequate for the task. And with targets restricted to French, Dutch, and Belgian sites, with many of its planes and crews drained off for the North African campaign, the Eighth was doing little in the fall of 1942 to prove or disprove the efficacy of daylight bombing.[5]

By that same fall, the Battle of the Atlantic—in which the Germans sought to isolate the British Isles and Russia by an all-out U-boat and air campaign against Allied shipping—was undergoing a change. U.S. shipyards were winning the production fight: November 1942 was the last month in which sinkings outstripped ship construction, and the Allies would soon gain command of the Atlantic by the use of small carriers, from which packs of fighter planes would hunt the U-boats.

It could be argued that the most crucial area of the war in the last months of 1942 was Russia. Thwarted at the gates of Moscow and denied the city of Leningrad, the Germans had reached the outskirts of another Russian prize, Stalingrad. Its capture would give the Germans the choice of driving north to encircle Moscow or south to the immensely valuable oil fields of the Caucasus. While opposing soldiers fought to the death in the rubble of the city, on November 19 the Red Army launched two counterattacks over ground sufficiently frozen to support the weight of their tanks, and in two days' time the German army was trapped; it would be ground up in the following weeks until its remnants surrendered.[6]

If the fall and winter months of 1942 saw a shift in the tides of battle, change and status quo were strangely mixed on the home front. Although the Selective Service Act was passed in 1940, not until October 1942 did Congress, under pressure from the armed services, lower the draft age from twenty to eighteen. Momentarily, some normally Democratic women became disaffected with the party which controlled the presidency and the Congress because their "babes (for all children remained babies to all mothers) were being sent out by Roosevelt to be killed in HIS war."[7] At the same time, many teenagers, previously exempt from military service, were tossed between emotional highs and lows by the looming prospect and abrupt reality of becoming soldiers, sailors, or marines. Increasingly, before their actual induction and reception, they became the targets of public scorn from men already in uniform. They must be slackers![8]

One thing that remained much the same yet reflected wartime influence was the American cinema. "Like the British, the American government counted upon the movies to contribute to national morale by supplying entertainment."[9] In such Wichita theaters as the Orpheum, Palace, and Civic, 1942 brought the traditional escapist fare of Cecil B. DeMille's adventure saga *Reap the Wild Wind,* with John Wayne among its stars, and the musical *Footlight Serenade,* starring the enlisted man's favorite pinup girl, Betty Grable. The war's influence was stamped upon several films of 1942, notably *Casablanca,* with Humphrey Bogart and Ingrid Bergman, one of the most enduring romantic dramas in American film history. There was also the recreation of an epic of early defeat which followed hard on Pearl Harbor, *Wake Island.* Bob Rea attended a movie in September called *This above All,* starring Tyrone Power and depicting the days of despair and hope following Dunkirk. Power, Robert Taylor, James Stewart, and Clark Gable were the most prominent of a contingent of Hollywood luminaries who joined various branches of the service in 1942 or 1943; one, Wayne Morris, became a Navy fighter ace.[10]

Movies were a year-round form of entertainment, but they were overshadowed in season by the hallowed American sports of baseball and football. Bob Rea was among the millions of Americans in uniform and out who huddled about their radios listening as the swift St. Louis Cardinals upset the mighty New York Yankees. It was as though men were not slaughtering one another across the oceans. (It would be the last season until the war was over that the Cardinals were swift and the Yankees mighty.) College football teams already had on their schedules service teams, such as Iowa Navy Pre-flight, that were manned with former college and professional players and were coached by such famed mentors, now in service, as Minnesota's Bernie Bierman and Indiana's Bo McMillan.

College football was involved at the end of the regular season in one of the most ironic tragedies in sports history. The Boston College Eagles, who were considered the Number 1 college team, suffered an overwhelming defeat on the afternoon of November 28, 1942, at the hands of traditional rival Holy Cross. The shattered Eagle players cancelled plans for a victory celebration at a popular downtown Boston nightclub, the Coconut Grove. Bob Rea's girlfriend, Phyllis Edwards, recently arrived from Wichita, was attending Boston University and that same night turned down a date to go to the Coconut Grove. Buck Jones, star of Western movies, was not so fortunate. He was among more than four hundred victims of the flash fire that destroyed the nightclub and the panic that jammed its exits with bodies.[11]

Nightclubs and hotel ballrooms were the profitable haunts of the Big Bands. Swing was in its heyday, with the groups led by Benny Goodman, Artie Shaw, the Dorsey brothers, and Glenn Miller selling most of the records and getting the fat contracts from Chesterfield and Old Gold, the cigarette advertisers who sponsored their weekly radio broadcasts. In the racially segregated society of the time, there was no such sponsorship for perhaps the finest of the Big Bands, those of Duke Ellington and Count Basie. By the fall of 1942, many of the popular musicians were being drafted or were joining up; the most successful of the band leaders, AAF Captain Glenn Miller, was working with the base band at Maxwell Field, Montgomery, Alabama. The great urban symphony orchestras still performed in 1942, but their younger male musicians were not exempt from the draft. In the Midwest, civic leaders proposed pooling the musical resources of the Kansas City Philharmonic and the Wichita Symphony (in which Bob Rea played violin) to keep the classics alive.[12]

*Life* magazine, in a November 1942 issue, published a photographic essay on the Middle West. One of the photos showed a farmer's barn at night backlighted by the glow from the nearby Boeing bomber plant at Wichita. The barn, *Life* informed its readers, would soon have to make way for an extension of the plant. Other photographs featured the main streets of several small towns in Kansas, Iowa, and Missouri. They gave the impression of stability and a sort of timelessness, as though depressions and wars might come and go, but these stores, theaters, and dusty intersections would remain unchanged. The economic importance of this land led *Life* to describe it as a cornucopia of foodstuff for domestic as well as overseas consumption. To illustrate its character, there was the mural by John Steuart Curry in which a farmer, pitchfork on his shoulder, advanced in step with GIs with bayonets fixed. Pulitzer prize-winning poet Carl Sandburg, with his look of craggy wisdom, was quoted as saying that "box-cars in the sky" were transforming the Middle West; the airplane was pointing toward a new world society. In another photograph, a Kansas family relaxed on their front porch on a Sunday afternoon. The father was head timekeeper in the division office of the Santa Fe Railway; the two sons were both in uniform—Army-enlisted khakis. In the swing, father read his Sunday paper and mother knitted.[13] It was a family, like most, not yet harshly beset by the necessity of sacrificing for the war, aside from contributing sons to the armed forces. Such a sacrifice was now widely shared, and many sons not yet in uniform might have echoed the feeling expressed by Bob Rea in a letter of December 23, 1942, to Phyllis Edwards: "I am not particularly enjoying life these days. . . . It sounds crazy, but I really wish the Navy would call me up right after Christmas."

Shortages, rationing, and inflation did exist in 1942. A nation on wheels scurried for recapped tires. Gasoline rationing was around the corner, but the Office of Price Administration set up a coupon system for the purchase of meat, coffee, shoes, and other consumer products. The cost of unrationed goods shot up as workers' incomes rose along with new defense plants.[14]

Another photograph in *Life*'s midwestern essay caught a group of students at the University of Iowa, Iowa City, lounging on the grass in front of Old Capitol in the sunlight of early fall. Some of the boys wore T-shirts; more than a few sported dress shirts and ties. The girls were clad in longish skirts and blouses. Their hair was long and full, the boys' was short and neatly groomed: no difficulty distinguishing between the sexes. But, said *Life*, "What the picture does not show is that Iowa U. is now a big training center for the Navy, with more than 1,500 pre-flight cadets in residence. Soon most of the boys in the picture will be in the Army or Navy."[15]

Between the fall of 1942 and June 1943, when Bob Rea began his formal cadet training, the Allied cause made significant progress in some theaters but dragged or remained in peril elsewhere. Grand Admiral Karl Doenitz, in charge of German submarine warfare, would later pinpoint May 1943 as the time when the U-boats' defeat became inevitable, thanks to a combination of aircraft and radar. While U-boats would continue to cause damage, they were no longer a major impediment to the bulk of lend-lease supplies that flowed into Russia, nor could they stop the thousands of American planes that were ferried to the Soviet Union via the ALSIB (Alaskan-Siberian) route. United States aid to the Russians was more generous than that extended to the British: "The British had to account for every item they received. The Russians received everything the United States could spare and the British convoys deliver." In spite of the massive losses the Soviets continued to suffer, the swelling influx of supplies and a four-to-one advantage in manpower enabled them to ready their own offensives after Stalingrad.[16]

The ocean voyage required for the build-up for a cross-Channel invasion became less and less hazardous in 1943, which allowed convoys and fast troop-carrying luxury liners such as the *Queen Mary* to increase noticeably the numbers of American servicemen and women in rural camps, villages, and cities of the United Kingdom. Supplies, from rolls of toilet paper to drums of gasoline and howitzer shells, took more and more space in warehouses and open air stockpiles covered with camouflage nets.

Daylight bombing of the Continent from bases in England intensified in 1943. The Americans did little to prove or disprove the merits of daytime bombing, but the Eighth Air Force sharpened its tactics, and as Churchill noted, "bombing only at night, . . . the British were at last succeeding in wrecking the Ruhr munitions centers."[17]

On the other side of the world, in February 1943, the Japanese suffered calamitous losses when a major effort to reinforce New Guinea was thwarted in the Battle of the Bismarck Sea, where AAF medium bombers skipped bombs off the ocean surface into crowded enemy transports, like children skipping rocks off a creek and into the opposite bank. In the Northern Pacific, the U.S. Navy put Army troops ashore on Attu in the Aleutians in May 1943 as part of a bloody but successful effort to drive the Japanese out of those barren islands and remove the enemy from Alaska.

One read about the far-flung theaters of war in such newspapers as the Wichita *Eagle* and the *Beacon.* The war tightened its noose upon colleges and their students after the fall of 1942. Bob Rea wrote to Phyllis Edwards on January 9, 1943, "The fact that fewer of us will be back next semester becomes more evident each day. K—— D—— and a freshman in V-5 . . . are leaving next week. O——'s K.C. trip netted the rather reliable information that all V-7s will be called up at the end of this semester. . . . I am still hoping to graduate. Am quite sure of it as [academic] credit is now being granted for partial [semester] attendance of men called into service." To get closer to the spirit of the times, as editor of the school paper, Rea launched a drive for compulsory physical training at Friends University, but, as he recalled years later, "That one didn't fly."

One effort that did not fly with American public opinion was the series of coal strikes in May and June 1943, whereby United Mine Workers boss, John L. Lewis, he of the bushy brows and biblical references, attempted to defy government restrictions on union activities and gain concessions from the ancient enemy, the mine owners. A public opinion poll suggested that almost 90 percent of the country viewed Lewis in a bad light; liberals and conservatives took turns denouncing him.[18]

Prophetic of future complications abroad was a series of *Life* articles dealing with a Middle Eastern problem that was older than this war and would outlast it. In the first of these pieces, the King of Saudi Arabia attacked the concept of a Jewish homeland in Palestine. He was challenged by one leading American Zionist but was seconded by another American Jew, who called for a "democratic, autonomous government" in which Jews, Moslems, and Christians would be equally represented. At the time of this debate, the Holocaust was in progress, its scale unknown in the United States.[19]

In April 1943, the lordly General Douglas MacArthur and crusty Admiral William "Bull" Halsey, the top commanders in the Southwest and South Pacific, agreed to meet; to the great surprise of their staffs, they got along amicably. Together they planned the next steps in the recovery of New Guinea and the remainder of the Solomons, a process that would extend into 1944. With increasing numbers of pilots trained since Pearl Harbor at either Corpus Christi or Pensacola, carrier- and land-based planes scoured the vast blue expanses of the Pacific in search of targets in the air, sea, and islands. Navy planes and AAF bombers and fighters from Port Moresby or Henderson Field preyed upon enemy airstrips and shipping, engaged the Zeroes, and flew ground support missions for troops struggling against the foe entrenched in the Green Hells below.[20]

While the tactical aspects of the Pacific war were being hammered out, in May 1943, the British bowed to American importuning for a specific date for a cross-channel invasion of Europe—May 1, 1944. Preparations began to accelerate, and several of the war's greatest air battles were fought in the following months. A low-level attack by AAF heavy bombers struck the well-defended Ploesti, Rumania, oil refineries on August 1. Hampered by navigational and command errors, relentlessly clawed by defensive fire, seared by the flames of their own targets, the American bombers paid a painful price for the damage they inflicted: fifty-seven B-24s went down, and in a few weeks the refineries were restored to nearly full productive capacity. In western Europe that humid July, U-boat facilities at the port of Hamburg became the focus of an RAF-AAF night-and-day coordinated attack. The RAF used a lethal mixture of incendiaries and high explosives that produced the first massive firestorm resulting from aerial bombardment. Close to 40,000 people suffocated or burned to death, a foretaste of other such phenomena, indeed a smaller version of the nuclear holocaust of the future.[21]

In mid-August 1943, the Eighth Air Force received the green light for a maximum effort against aircraft manufacturing plants and related industries. Constantly harried by deadly German fighters, the Eighth Air Force aimed at destroying the most serious threat to daytime bombing, the Nazi fighter-interceptor, at its source. But before August 17 was over, sixty smoking bomber wrecks littered the European countryside, most of them brought down by the "nightmare fighters."[22]

Although Americans did not share the intensity of British interest in the Mediterranean, they realized the importance of the island of Sicily as an Axis base for attacks on Allied convoys. Early in July 1943, an invasion armada, larger than that which would invade Normandy, disgorged men and vehicles on the southern beaches of Sicily in the face of light opposition. The operation was embarrassed when jittery gunners of the offshore fleet mistakenly fired upon U.S. paratroop-bearing transport planes, causing high casualties in this first major Allied airborne operation. Along with the clash of arms there was a clash of egos between the two ranking ground commanders, Montgomery and Patton. Patton beat his rival to the key town of Messina, but his showboating and Montgomery's tardiness allowed the enemy to evacuate most of their forces across the narrow Straits of Messina to the Italian mainland. The most notable casualty of the Sicilian campaign was Benito Mussolini, Il Duce, whose own government forced him to resign while the invasion was in progress.[23]

On the American home front a disturbing series of events during the hot summer of 1943 spotlighted a basic contradiction in national attitudes that most Americans embraced unwittingly. They had embarked upon a crusade to rid the world of Nazism, with its extreme racist philosophy, but in reaction to Pearl Harbor they had come to view the Japanese, those little yellow people, with a hatred and contempt that had strong racist overtones. In this spirit they had locked away many Japanese-Americans in the democratic version of concentration camps, far milder places than the Nazi camps but nonetheless places where the government blatantly denied basic human and constitutional rights. The violence underlying American racism erupted in July and August in riots in Detroit and Harlem. *Life* magazine caught the implicit contradiction in American society that summer. It proudly reported the cruise of the Coast Guard cutter *Campbell,* on convoy duty in the Atlantic, and described how ship's gunners, including a "Negro crew that . . . fired a steady box of 30 rounds," sank a Nazi U-boat. In the same issue, *Life* ran a series of photographs of white mobs savagely beating lone blacks and the picture of a white policeman, staggered by a well-aimed bottle while trying to protect the victims. Some of the whites, reacting against the police presence, had "raised the cry of 'Hitlerism . . . Just like Germany . . . You damn gestapo.' " *Life* commented, "They did not realize they looked and acted like Nazis."[24]

These portentous episodes of the long, hot summer of 1943 made little impression on servicemen embarking upon their basic training. Among them was Bob Rea.

# Naval Flight Preparatory School, William Jewell College, Liberty, Missouri, June 9–August 30, 1943

*Aviation Cadet Robert R. Rea was ordered to report to the Naval Flight Preparatory School at William Jewell College, Liberty, Missouri, on June 9,*

George and Fleda Rea—Bob Rea's "Dear Folks"

*1943. Rail transportation was provided from Wichita to Kansas City. Having the afternoon free in Kansas City, he wrote a postcard home:*

June 9, 1943

Dear Folks,

So far O.K. Spent afternoon in the [William Rockhill Nelson] art museum—very delightful, hope to get back some weekend. Hot as Kansas here and twice as sultry.

June 12, 1943

Dear Folks,

This is just to let you know that the scion of the Rea family is still alive and kicking. I should have written long ago, I suppose, but this is the first opportunity to really relax since we arrived, also just got a return address.

I suppose I should say that the trip up was rather enjoyable. I spent a rare and wonderful (so it now seems) hour and a half in the Rockhill Nelson Art Museum. It is a beautiful and interesting as well as educational place. If and when I get back to K.C. I shall certainly revisit it. The cultural, cool, refined, and peaceful atmosphere seems like a dream compared to this place.

Which logically leads to a description of the station. Liberty, pop. 4000 including cadets. Wm. Jewell a Baptist college well-endowed and equipped. Location comparable to Southwestern [University, Winfield, Kansas]—hill. All cadets live in barracks according to battalions. We are the 6th [Battalion] app[roximately] 150 men. Live 9 to a bunk room, double-decker bunks. It is hot and extremely sultry here, rain every other day. Most pleasant surprise is the chow. It is plenteous and good, no style but quality and quantity. Haven't weighed, but should be gaining a pound a day. No uniforms yet, but prospects of khaki. Wearing shirts and ties constantly. Work so far is indoctrination—and drill—really quite enjoyable except for loss of shoe leather. Studies begin Mon[day].

There is a great deal more that I might write about, but will save for tomorrow's lengthy epistle. Now, I want 1. a letter, 2. a package containing the following—notebook on desk, stationery, AKT "t" shirt, khaki work pants in closet, best brown shoes.

June 13, 1943

Dear Phyl,

... I am now at Naval Flight Prep School at Liberty, Mo., pop. 4000, temp[erature] 90°, humidity high. The place is really quite enjoyable as long as you have no time to think—and the Navy sees to it that you don't. We have about 600 cadets, few uniforms, wonderful and plenteous food, marching, and regimentation. The whole trick to life here is to forget you were ever Bob Rea or ever knew anything or anyone, ever thought or imagined anything, and just start living all over. I just hope I come out with some sort of personality left. It seems 4 years instead of 4 days. Our day begins at 5:50 and ends at 9:50! Quite a change. Am living in former women's dorm—9 men to a room. Quite homey, oh yeah. One good thing: life for the next three years will make me realize the great pleasure and joy of civilian life in peace time. All officers and men here are working and hoping for a return to that situation.

June 13, 1943

Dear Folks,

... Sunday schedule includes church at 1000 (I am in choir which is awful). Went to Liberty Christian Church at 1100, rather nice little church. For a sample, here's what we had today: fruit cocktail, 3 pork chops, potatoes and gravy, green peas, 2 hot rolls, butter (all you want), milk, 3 cookies, ice-cream. As I said, the heat is terrific, thanks to humidity. Oh, yes, I can prove I'm in the navy by my haircut now—½" all over. For your benefit in deciphering these letters (written in my bunk), bunk is bed, deck is floor, bulk-head is wall, chow is food, 6th Batt are those who entered same time I did, Melrose Hall is our barracks.

So—here's the deal. This is strictly Navy—and that means real

Phyllis Jeanne Edwards—Bob Rea's "Dear Phyl"

discipline. We march and muster a great deal and whenever we go any place en masse.

Our week of indoctrination included government insurance $10,000, made out to Mr. G. E. Rea, free too. Then we are all urged to buy war bonds, purely voluntary of course, with firm recommendation that we purchase. So you should receive $18.75 bond per month while I'm in training. Please hold them with the other bond. We have been seeing dull Navy shows in the evenings, drilling in spare hours, eating too much, sleeping hard, and wondering when the real work begins.

We have a pretty good bunch of men, largely college men of 2 or 3 years. Also a large contingent of enlisted seamen here for officers' training. They are all about 25 and fine men, physically. The general intellectual level leaves much to be desired, but there are a number of fellows who should make pretty good shipmates.

For living quarters we have a large corner room bunking 9 men. One small closet. Shower and bath facilities two floors down. Not quite like home, and I don't know how we'll all manage to study.

Our morning schedule begins at 0550 or 5:50 and ends at 2150 or 9:50, which seems short but also plenty long. Textbooks issued so far seem simple enough.

It seems we are to be here for 12 weeks if we pass all tests on time— otherwise 16 weeks. At present, this seems like not too long to study, but plenty long to be here. One thing is definite—this turn in the Navy makes one appreciate the pleasures of civilian life in peace time. It is interesting to note that in this most specialized and select service, all, both officers and men are ardently working and hoping that the return will be soon.

Please write often, length is not essential.

June 16, 1943

Dear Folks,

A few minutes off before chow, so I'll pen a line. We are off on our studies now, and so far so good. The math, first hour, is very simple review and goes easily. Next two hours are devoted to PT (physical training). We have had tests (push-ups, chinning, jumping, indoor obstacle [course], etc). Fourth hour we are back in math, this time for blackboard work. After chow, we have communications. Have already learned the Morse code and begun on semaphore. Don't know yet, but it doesn't look too tough. Next—recognition of air and sea craft. This looks pretty stiff. We have to learn to recognize 40 different planes and ships, any angle, at 1/75 of a second. Navigation follows with two hours of very dull stuff. So far nothing but geography—and the prof is awful. Can't speak English, correctly, that is, and apparently knows nothing of his subject. We are certainly hoping he will know something about actual navigation. The communications course is taught by phonograph record and motion picture, no lecture at all. At present we are in "study hall," confined to our rooms. You can imagine what it is like with 9 of us trying to study four different courses, talk, ask questions, practice code, and keep quiet. Concentration is *really* important. There's really little news around here. Found a pair of foils and masks this a.m., messed around for two minutes with our platoon leader, a tough ex-sailor-Italian from Chicago. He was willing, but not able, but it was fun. The town has a very nice U.S.O., former bank building, and it is almost exclusively for cadets. Have met several congenial fellows: one from Southwestern who knows the W——s, another from Coffeyville who swam against Jack W—— in high school. Am planning to go to K.C. about next weekend to re-civilize myself. It isn't so bad though if you're the first up, first in the shower, and learn when to be where. We get downtown privileges 1700–1830 (5:00–6:30) every day, and overnight Sat[urday] and Sun[day] once a month. . . .

. . . If you have not sent my white shirts already, don't do so. This is no place to wear white shirts—on second thought, you might send the thin white one. As I need shirts (other than the old ones I have) I will buy cheap brown ones until uniforms are issued. I may need the old green pants—but not yet. And send the raincoat. This is the wettest place I ever hope to see—and we are as far from every place as possible: two blocks to mess hall, four to classes. . . .

June 17, 1943

Dear Dad,

Your letter came today and also the package, although I haven't picked it up yet. Glad to hear all the details of normal life—even the grief of the business world. Keep it up.

Not much new around the N.F.P.S. today. The river, "old Missou," is very high, and all battalions except ours have done duty on the levy today. We are looking forward to a call tonight. Communications proceeds but slowly; we are now into semaphore before polishing off buzzer. Last night I stood watch for the first time, 12:00–2:20. Very dull, and really cuts the night in small, small parts.

. . . We're getting into planes now in Recognition and talk flies thick and fast concerning planes, engines, etc.

A word about my room-mates. Three ex-sailors, P——, P——, and R——. P—— is a nice blond kid from Chicago, quiet and clean. P—— is a second generation Austrian. Clean built fellow who goes in for gymnastics. He is one of the finest fellows in the room, but weak on his studies. R—— is a Tennesseean with a nice soft accent. Oldest man in battalion, he just got in under the age limit, 27. Then there's P—— and P—— from Springfield, Mo. Just a couple of college guys. R—— from St. Joe, Mo., R—— a farm boy, also Missourian, and P—— from Wichita. A very average and representative group.

June 21, 1943

Dear Folks,
. . . Today the 6th Bat was put into T-shirts for the first time—and we really welcomed the change. Of course we didn't make the change until after P.T. Marched for two hours, got good and sweaty and dusty, or should I say muddy, and then the L[ieutenan]t. yells, "Six minutes till classes—get going." No shower, scarcely time to dress. Much grumbling about officers. The two younger athletic officers, J—— and K——, are known as the "torture twins." First week grades came out today. I was rather well satisfied. They are—Math 3.4 (tsk, tsk, tsk), Communications 4.0, Recognition 3.8, Navigation 3.7. Bearing in mind that 4.0 is perfect, my average is at least passing. Got a 4.0 on my Nav. final, or rather weekly, but some low dailies pulled it down. . . .

June 29, 1943

Dear Folks,
Tuesday morning and it's nice and cool. Slept under blanket last night! Last week my average stayed at 3.7. Our platoon, 6B2, ranked 4th in the regiment with a 3.4 average. Not bad. Navigation is getting tougher. We're doing plotting now—and it takes time. Also starting Blinker in Com[munications] and that's not a push-over. P.T. now consists of about 40 minutes of basketball, exercises and relays, and same amount of volleyball. Good exercise and plenty of it.
About my K.C. jaunt. Had no intentions of going in 'til Sun. morning. It was a perfect (and hot) day, so three of us hopped on a bus and away we went—standing up. Round trip is only 66¢ so it's hardly even a luxury. Arrived in K.C. at 11:00. . . .
. . . We had dinner at the Muehlebach Hotel Coffee Shop. Very nice—but no nicer than the Lassen [Hotel, Wichita] and food not as

good. We debated what to do after dinner and decided on Swope Park—my companions not being the "arty" type. It was a good choice. It was beautiful out in the park. Lots of people and animals to watch and popsicles to eat. On the way back we stopped off at Pla-moor, the big recreation center—dancing, bowling, billiards, swimming, roller-skating. Back home on the 6:30 bus. Spent the evening studying, I think.

Yesterday we received the glad tidings that there would be no review Sunday. Today is a red-letter day, due to our first shots—inoculations for tetanus, typhoid, and smallpox. More sore arms in the battalion than sunburn. At present everyone has a good headache and fever and is ready to call it a day.

This navigation plotting is really interesting stuff, but takes a lot more time than we ordinarily have. The schedule for study for a while seems to exclude all else.

July 4, 1943

Dear Phyl,

I am writing to you (obvious fact) because: 1. It is hot and sticky tonight, 2. It is the 4th of July and Liberty is awfully dull, 3. I am thru studying and letter-writing seems in order, 4. I feel like writing to you.

(July 5) As I was saying—just drilled for two hours in a broiling sun. x#'&*! We're having regimental parade this afternoon, and everybody is in a foul mood—naturally. Got uniforms last week. Regular army summer khaki, over-seas cap with Navy emblem, black tie and black shoes. Not particularly flashy, but lots better than civvies. . . .

The 4th was historically quiet—spent the day resting and writing letters, and going to a show I'd already seen once. Very dull. . . . Speaking of sewing, I'm becoming very proficient at shortening trouser legs—still can't put a button back on. Some day I must learn how to do that.

(July 6) Will I ever get this finished? The review yesterday was a blistering success. Very impressive though.

July 6, 1943

[2 postcards]
Dear Folks—Having just talked to you, there isn't much to say, so I'll send you the best available photos of N.F.P.S. Dad will be happy to know

I made a 4.0 in math this week. Navigation is getting much tougher but more interesting. Plotting all the time now. Uniform issue consists of pants, 8 shirts, 3 caps, 6 pr. socks, 2 pr. shoes (1 high, 1 low black), 6 skivvy or T shirts and 6 shorts, 2 belts, much stuff like work hats, gloves, ties, etc. Very quiet 4th—hardly any fireworks. First "semester" over next week—Finals. Brr. Write.

July 7, 1943

Dear Folks,

. . . Our celebration of the 4th on the 5th was a dusty success. Many compliments on our marching—6B2 shown as a pretty decent drill squad. Afterward we all saw a free show at the local theater, courtesy farmers whose land was saved by the cadets during the flood.

My weekly average for the last week was 3.75. As I said, I cracked a 4.0 in math, but slipped to a 3.3 in recognition. This week we are taking trigonometry in math. New to me, but I seem to be getting it pretty well. Final this Saturday, then we start on Physics. Navigation is nice and hard, but very interesting. We can now plot fixes on radio bearings, calculating wind drift. Great stuff, but time-taking.

Attended my first meeting of the *Fledgling* [newspaper] staff last night. Will prove very interesting and profitable, I think. Today Lt. Bice, O-in-C, announced special privileges for choir members. We must attend regularly, but are granted overnight liberty every three weeks instead of four. We may be out till one o'clock Sat., and when on liberty don't have to be back till 7:00 Mon. morning. That means liberty from 5:15 Sat. p.m. till 7:00 Mon. a.m. and should make it possible for me to get home. . . .

Now about my clothes. I think I sent a G.I. skivvy and shorts home. Please return same, clean, in next laundry bag. No great hurry. . . . I'll probably be sending shirts and pants home to be cleaned. Would you like to shorten four pair of trousers for me? I don't relish the idea one bit. Anything else I send home, clean or dirty (that is, civvies) is to stay. . . .

Incidentally, we are definitely here for sixteen weeks (that is, 12 more).

July 9, 1943

Dear Folks,

Hurray! For the first time since I got here I have spare time. Of course I could be studying recognition, but who wants to when you can rest. The Admiral (Young) came today at 1430. Everybody dressed up, choir sang "Anchors Aweigh." The Admiral made a nice short speech. 6B2 drilled badly in a hot sun on slick grass. The rest of the regiment also baked. The Admiral left. Regiment secure for the day. Three cheers. Nothing to do from 1500–2145.

Have my math final tomorrow. Today I made a 4.0 on geometry and trig, making my math average 3.7. And makes me feel pretty good.

Admiral Young (Rear Admiral) is Head of Bureau of Supplies and Accounts, and Paymaster-General of the U.S. Navy. For story of his visit, see July 10 K.C. *Star.* Also had pictures of 6B2 taken, but doubt if it gets in.

The cookies came and went—almost in one breath. Boy, oh boy, were they good!!! Everyone sends thanks and request for more. . . .

July 13, 1943

Dear Folks,

A slight explanation of my plans. My special choir leave was arbitrarily set for this weekend. It couldn't be changed and so—feeling the urge to travel—I decided to make the trip. One Elmer B. M——, of Winfield, a very nice fellow, is traveling with me.

. . . As time is a bit limited, we might as well skip a few hours of sleep—reveille at 7:00. And breakfast—coffee, not cocoa; no cereal or breakfast food; lots of fruit—and no oranges; preferably, no bacon either. You can guess what I've been eating. Dinner—and lots of it, especially iced tea. And have a little gas in the car, Dad, please.

Had a good Sunday p.m. in K.C. at the Art Museum. My weekly average was 3.9. Final Math grade 3.8. Now studying Physics and Principles of Flight. Kind of tough. . . .

July 16, 1943

Dear Phyl,
   Your letter came today—along with six others—to make this a red-letter day. I am a bit high for a naval air cadet—I am going home tomorrow night—for twelve hours. Don't have much studying to do, so I'll try to make this a decent letter.
   Five weeks gone—ten more to go. It seems we've only been here a short time. Oh, yes, we got paid today, too, $90.00! No wonder I feel good. So far so good on the academics—3.9 average last week. However I fear this week won't be as good. We are having Physics now instead of Math—and it is about twice as hard. Navigation is really getting complicated, and very interesting at the same time. So it goes—much faster than school ever went. Saturday seems far away on Monday, and today [Friday] Monday seems yesterday. In a way it's wonderful—keeps you from going mad. . . .
   This Navy life ages fellows fast—even this early in the game. You can see it in the new men, and our Batt, and those who have been here three months.

July 25, 1943

Dear Phyl,
   I was very glad to receive your letter. Mail call is the only thing that can match chow call for popularity. I spotted your blue envelope long before my name was called. Ha, says I, that's from Boston, and lo and behold, it was. . . .
   There is no news from Liberty. The place is not conducive to high intellectual endeavor. I'm keeping my grades up pretty well, haven't killed myself on the physical training, and am getting a swell tan from 2 hours of drilling each day.

July 26, 1943

Dear Folks,
   Something tells me I am a bit behind in my correspondence. Hm, could it be I've been busy? And how. Last week was awful. Got a day

behind in Navigation and like to never caught up. No time to study. Missed a Physics class for choir, flunked two daily quizzes in Nav[igation] and did badly in Recognition. They threw six words a minute in mixed letters and numbers instead of plain language. And to top it off, I was room captain and did most of the cleaning each morning and noon. However—there's always a silver lining. Knocked off 4.0 in Physics, Communications, and Navigation, 3.9 in Rec., 3.8 in Flight. Which wasn't too bad, was it.

Saturday was a red-letter day for mail, three packages. Clean clothes, new shirt, and cookies from the Delta Rho's! The clean clothes were very welcome. You should get my laundry case with this mail. . . .

We are now living a semi-civilized life. Got ten lockers for our room and no longer have to live out of our bags. A very pleasant change, believe me. . . .

Had our first day of military track and football. Slightly on the rugged side, but at least it will help condition me. Should have a beautiful tan next time I get home. Guess that's about all the news, so I'll get busy on my navigation. Oh, yes. The choir sang at the Presbyterian Church Sunday. Greatly appreciated. The minister referred to the "need for the Christian mysticism of that group led by Rufus M. Jones" in modern religion. Ha! The Quakers come thru even in Liberty. Write soon.

July 27, 1943

Dear Phyl,

. . . Last week was *very* hectic (might even say hellish) but totalled up pretty well. So, in contrast this week is going along first rate. We're in track and football now. Run a few laps, run a few dashes, run the obstacle course—and you're all warmed up for a fast game of football. It's warm at noon, too. The latest scuttlebutt has it that we may get an extra liberty soon. Have had a number of days of perfect inspection, and the gold-braid really likes that stuff—are we hoping? And how!

Got a box of home-made cookies from the Delta Rhos last week and a group letter yesterday. Both were swell. I used to smile at the DPAN [Delta Rho Alpha Nu sorority] "Boxes for the Boys," but they are really swell. And do they hit the spot? Yes.

Chow time, so I better dress (shirt and tie). Write soon.

July 29, 1943

Dear Phyl,

On watch again—midnight, and I am off in half an hour. . . . At night when the barracks are quiet—the only time, I can relax physically and catch up with myself. It's the same feeling I get Sundays when I go into K.C. and visit the art museum. There is a Spanish court—fountain—green grass below, blue sky above, rising and falling water—white and green, faded colors of old painting, and symmetry. I can sit for hours just looking and seeing nothing, yet visiting every place my mind may wander.

August 2, 1943

Dear Folks,

At last, at 2200, twenty hot minutes to write a letter. I've been on watch since chow time so am behind on both studies and letter-writing. Too tired to study, so I'll write.

About the plane trip. Superlatives are out—too tame. The plane itself was swell; nice soft seat right over the leading edge of the wing—a perfect seat. The curtains are pulled during take-off and landing, but you can see enough even then. There is a slight feeling of speed on the take-off—you really feel those old motors pulling you along. Then, tail up and in a minute you get a change in perspective and you're airborne. It is wonderful to see the ground drop away. We lost no time in squaring away for K.C. We flew at cloud height—right thru them—and when you talk about blind flying, you mean blind. The trip was quite smooth, but by the pressure against the seat, you can tell when you hit an updraft or drop into a pocket.

The country looks beautiful. Checkerboard pattern in blues and greens with a plowed field here and there. You can almost check your course, plot winds, from the roads. The clouds were beautiful from above and cast numerous shadows on the fields. Strictly according to Navy navigation texts, these shadows, even at low altitudes appear to be wooded areas. Rivers show up wonderfully, especially the Kaw. The hostess was quite charming, inquiring whether I wanted anything to read or a glass of lemonade. I was, however, much too busy enjoying myself to bother with such details.

Then, in about the time it has taken to write this, or so it seemed, we are over K.C. and banking in at Fairfax Field and it's all over. Back to Liberty on an early bus and a good night's sleep.

If I ever had any doubts about flying, they are gone now. Any amount of mental effort or physical punishment will be well worth it if I get those wings at the end. In any event, it has been worth it and will be as long as I'm in the air corps.

If you get a chance to fly somewhere, do, by all means. You'll never spend a more enjoyable hour and ten minutes than I did yesterday. The trip home seems like a dream—a very swell one, but too short to be real. Now I can imagine I only dreamed of going home and fell asleep daydreaming. I don't need to say I'll be home again in three weeks. . . .

August 3, 1943

Dear Folks,

Well, this seems to be a regular thing; last week was easy, this week hard. Had a good 3.9 average last week. Perfect 4.0 in navigation. This week the semi-final test, which promises to be bad. We are running in track. Yesterday it was 7 laps, today 3, and a half a dozen 100 yd. dashes. Rugged.

We're starting on ships this week in Rec[ognition]. Very interesting, too. Have a final in Physics, too, so I'll be ready to come home. I won't know what train I'll take until Saturday, so just leave the screen unlocked. I'm trying to get out early and if so will get in around 10:00; if not will arrive about 1:30 but don't know schedule as it has been changed. Will leave home on the 5:20 train. So don't wait up, and leave the screen unlocked and I'll be in as soon as I can.

Enclosed is "my" first *Fledgling*. The front page is mine except for the cut. Wrote most of the heads and more than my share of the stories— 25%. . . . Don't claim any part of the proof reading. Almost as bad as the [*University*] *Life!*

August 11, 1943

Dear Folks,

Well, at it again. And how. Engines and Aerology promise to be plenty tough. Navigation is off on a new and more interesting phase.

Here's the news for the week—they are fumigating the chow hall and galley this weekend and granting liberty from 3:30 till midnight Sunday. Lots of time to come home on the streamliner, get in at 9:20 and go back the same way—and I have a watch! So—I won't be home and am going to Lawrence instead to visit Gailen W—— and Charley B——, spend afternoon in Art Museum in K.C., and leisurely back to Liberty. However, I'm glad I had last week. I'll never see J—— and B—— and H—— again for a long, long time—if ever. I took my Wednesday night Liberty liberty tonight and had a nice meal and break from routine. Very nice. The clothes came through today, and I'll return the case about the end of the week. Thanks a lot.

The trip home was something to write home about. My train was supposed to leave at 5:40 and arrive K.C. at 11:00, as you know. I wanted the 12:15 bus back to Liberty. Well, as you also know, it left about 6:05. It was very crowded and went slowly to Newton. There we stopped for more passengers and ice. The water was hot—and they ran out of paper cups. Well, we sat in Newton until 8:00—just sweating. And then we really travelled. Made Kansas City about midnight. When I got into the Union Station waiting room it was 12:10 by the big clock. Well, where there's a Rea there's a way, so I tore outside and made the first cab up to the Pickwick—clear across town—and we had to stop for every red light. Paid off the cabby and jumped out the door of the cab just as the last Liberty bus backed out of its stall. Whew. Made it. Even with five hours sleep I slept through every class the next day. By now I'm rested up. . . .

We're having quite a mystery this week. Some female in Wichita is sending me post cards with good wishes and guess who—and some clue as to who it is, like, brown hair, blue eyes, etc. So far I'm pretty well baffled. Two more days and I find out who it is.

The *Fledgling* remains to be done this week. Tonight, when we should have done all the work, the editor took out on liberty and left no instructions, copy, typewriters, or encouragement. Fine stuff. We'll end that when I take over.

August 13, 1943

Dear Folks,

Not much news, but I have so much to do that I might as well write a letter. Being Friday 13th, I fouled up a navigation quiz completely and ruined a 3.9 weekly average. Mentally chastising myself, I walked down to the print shop to see how the *Fledgling* was coming along and met Ens. Brown who offered his congratulations on my being 4th ranking cadet in the 6th Batt. I thought he was joking, but no, it's true, and I have a 3.8 average. Also, I may get out with the 5th Batt, but hope not as I'd miss a month of studies—but to get out and fly—ah! However, that remains to be seen. I am extremely jealous of everyone going home this weekend, but am looking forward to a nice visit and afternoon in K.C. We got treated for trench mouth today, as several cases have appeared. I don't have it, and the treatment is most unpleasant. Not much else so I'll close and get busy at something more nautical.

August 16, 1943

Dear Folks,

Another week gone, and we're back at it. Had a very delightful weekend in Lawrence with Gailen and Jeanne W——. They have a very nice little trailer up on the Hill. Most restful cool night, and we slept in till 9:00. Sheer luxury. Jeanne cooked breakfast, which was very good, and then I went over and saw Charley B——. Both Gailen and Charley find medics pretty stiff stuff. All work and no play. Charley doesn't like the Army and is ready to get out as soon as the war is over. We all had dinner at the Eldridge Coffee Shop, and I caught the Streamliner in to K.C. Went out to the Museum, a show, and very good dinner at the Forum, big cafeteria. Chow around here has been big thick sandwiches and milk the last couple of days; however, the mess hall is now operating and we are getting back to the old schedule. I take over the *Fledgling* this week for one issue at least. . . .

August 22, 1943

Dear Folks,

Officially reporting on activities of A/C R. R. Rea. Finishing a rather rough week with a couple of rougher tests. Went into K.C. with a friend and got a nice air-conditioned room at the Pickwick. Had dinner with two girls who graduated from Friends this year. We had a swell time, and dinner, talking over old times and catching up on the latest gossip.

They came up last week and are studying meteorology to be Junior Weather Forecasters or something. Civil Service job. They will be in K.C. for five more weeks, and two more are coming up this week. We should have some swell times before I leave. Slept until 10:00, bathed, shaved, and had waffles and sausage for breakfast. Loafed around the Canteen for a while and then went out to an amusement park and rode the roller-coaster. Fine stuff. After an hour or so we went up to the Liberty Memorial and enjoyed the view and a recital on the Hammond organ. Very nice and restful. . . . Laundry case and cookies came very timely. I was completely out. Am returning it as usual. . . .

August 23, 1943

Dear Phyl,

. . . News of P—— and B—— is welcome. Lucky devils to get their commissions so soon. Five months—and in the same time I'll just be beginning to fly, not even within sight of a Naval Air Station. However, I wouldn't trade with them when I get my commission. And believe me, when it comes to rating with the enlisted men, you can't touch the air corps. But enough, I'll be a very lucky fellow if I make it. . . .

# 3

## Sprouting Wings

THE FALL OF 1943 found Bob Rea at the University of Kansas War Training School, where he received his first instruction in flying. Other Americans were learning harsher lessons in the school of war at Salerno, Schweinfurt, and Tarawa.

Montgomery's British Eighth Army kicked off the invasion of Italy on September 3 with a thrust across the Straits of Messina; the U.S. Fifth Army, under General Mark Clark, came ashore at Salerno on September 9. Montgomery's landing was unopposed, but the Americans met fierce resistance from a powerful German force determined to make Italy a battleground, even though the Italians had thrown in the towel. In late September the British captured several key airfields that made close support easier, and in October Clark's Fifth Army slugged its way into Naples and gained a port through which essential supplies could be funneled. Allied priorities, however, were higher elsewhere; Italy was and would remain "a secondary front, which was subject to all the frustrations of such an operation."[1]

On October 14, 1943, a Thursday, the U.S. Eighth Air Force in Britain sent all its available B-17 Flying Fortresses on a return visit to Schweinfurt in an effort to knock out the ball-bearing *werke*. Climaxing a series of already costly deep-penetration missions in September and October, "Second Schweinfurt" again found the bombers flying unescorted during much of their journey over German terrain. Once again the Luftwaffe fighters swarmed. The mission turned into "Black Thurs-

[ 79 ]

day" as U.S. bombers dropped in sweeping arcs, spun in, or exploded in the sky: sixty of them before it was over.[2] To the critics of daylight bombing, it seemed the last straw. Indeed, it marked the virtual end of the AAF's stubborn self-destructive effort to demonstrate the viability of pinpoint bombing through the use of the "self-defending" heavy bomber and its defensive formation, the "Combat Box." At the end of October, General Henry "Hap" Arnold, commander of the AAF, ordered that the Mustang, a newly developed long-range fighter, should be assigned exclusively for the protection of the Eighth Air Force bombers.[3]

Even while the shock waves of "Black Thursday" were reverberating between Europe and the United States, on November 11 a mighty armada of ships and men sailed for an atoll in the Gilberts, a flyspeck on the rim of the Japanese island defenses, one of many which were strung out like rows of barbed wire to protect the Home Islands. The name of the objective of this amphibious operation—Tarawa—meant little to the thousands of Marines crowded aboard the transports, nor were they aware that they were inaugurating a new type of warfare to be known as "Island Hopping." Their attack was preceded by two days of naval gunfire and carrier air strikes designed to break the enemy's will and destroy his ability to resist, but when the Marines headed ashore at Betio, machine-gun fire whipped the water around their "Amphtrac" carriers and turned it into a bloody froth. From bunkers and pillboxes, the Japanese lashed the invaders with a steady hail of shell and small arms fire. Banzai attacks carried to the water's edge, where corpses clustered like seaweed on the beach, bobbed in the shallows, and hung on the exposed reef. It was an inauspicious beginning. Slowly, with a grim determination that matched the suicidal Japanese resistance, using flamethrowers and explosive charges to reduce the bunkers and pillboxes, the Marines moved inland. It took three days to secure the island and cost over 1,000 leathernecks' and sailors' lives. Some lessons were learned, among them the merits of a new U.S. Navy fighter. A formation of F6F Grumman Hellcats engaged a superior number of Japanese aircraft, including Zeroes, over Tarawa and shot them all down.[4]

Many of the ships conveying Marines and GIs with their new semiautomatic Garand rifles to the Gilbert atolls were recent products of a new American industrial revolution. "The Furnaces Roar," proclaimed *Life* in September 1943. Pictures and text documented unprecedented strides in essential components of wartime industry: merchant ships from Kaiser and landing craft from Higgins; steel whose output promised to exceed that of the rest of the world; trains "carrying more traffic of every kind than has ever been carried in any country at any time."[5] While celebrating the marvels of the "Arsenal of Democracy," *Life* pointedly contrasted a somber vignette of war with a seasonal celebration of the civilian life that Bob Rea was observing in Lawrence, Kansas. The assistant adjutant of a heavy bomber squadron was shown sorting out the personal effects of a missing bomber pilot, which were to be boxed and shipped home to his wife, or widow. The following article featured models displaying the latest college coed styles for the school year of 1943–1944. The fullness of the new fashions scarcely suggested the pinch of war, although some of the perky headwear resembled that worn by a growing number of servicewomen.[6]

As one of the top men in his class, A/C Rea left Flight Prep early and with little warning, but assignment to the CAA-WTS unit at Lawrence was a real plum. There Bob took to the air. The excitement he felt comes through in his letters, but there was little time for reflection between one challenging maneuver and the next. Retrospectively, he should have mentioned the creaking and cracking and banging noises that characterized the little Piper Cub, the difficulty of seeing around the hulking shoulders of a fat, six-foot instructor in the front seat, the gustatory pleasure of apple pie à la mode and steaming hot coffee after a cold morning flight—almost a ritual.

Toward the end of this period, a new note began to appear in his correspondence (even though the faithful laundry case continued to make its weekly trips home to Wichita). Rea knew that he stood academically with the best of the competition, even if he could not run on the 440 relay team or clamber over a sixteen-foot wall with agility (rather more like agony). Proving to himself that he could indeed fly was a great personal satisfaction. But if flight was exhilarating, it also had a sobering effect. The change is easier to appreciate than to express; the ground never feels quite the same once you have left it.

Bob Rea in WTS (War Training School) cadet
uniform with overcoat—green garrison cap with V-5
emblem, khaki shirt with anchor collar pins

# War Training School, University of Kansas, Lawrence, Kansas, August 30–November 17, 1943

September 5, 1943

Dear Folks,

. . . I suppose I should start back at Monday of this last week [August 30]. Monday morning we received instruction on the use of plotting boards and that afternoon took the Navigation final. It was tough and only gave me a 3.2. Tuesday we did some more navigating and a lot of high-class goldbricking. That evening at 4:00 they told P——, another of my roommates, to be ready to go to Nebraska at 5:00. We rushed around and got him off, and then P—— and R—— and I went downtown for dinner and a show. They left at midnight. Our bunch was ushered out of NFPS with a chicken dinner—and then waited an hour and a half on a train. It was the oldest rolling stock I ever saw. Red plush seats and coal oil lamps, burning. We left K.C. immediately for Lawrence and were soon in our temporary quarters in the TKE house. Very nice frat house. Two doors from where Charlie B—— lives. The first day was the toughest scholastically I ever went through. Four hours of navigation, two of aerology, one of code, one of C.A.A. [Civil Aeronautics Authority] Regulations. We got entrance exams in all subjects. I had been taking code at 10 words/min. at Liberty, and here we were handed 14 w.p.m. Tough! Nav. was bad, I thought, but despite my month less [of] training, I had the highest test paper of the 30 men sent here.

Here's the setup. Reveille at 0600, breakfast at 0630, nothing but study or sleep or loaf until 1000. Then catch a bus for airport and fly, returning at 1200. Afternoon classes at 1300 to 1800, taps at 2200. One night a week off until 2300, weekends from secure or chow until Sunday 2300. The two officers on the base are fine fellows. Discipline is practically nil. Wear full uniform, black shoes and tie all the time. On liberty, we can wear greens, although they are a bit hot at present. Saturday morning I had my first hop. It was great. The Piper Cubs we fly aren't much to look at, but they do the job. All I did was 90° turns, climb and glide, and taxi. My instructor said it was too mechanical and my stick and rudder (hand and foot) coordination wasn't perfect the first time, but I was quite satisfied. As I remember, I didn't shift gears perfectly the first time, either. The last time we tried it he admitted it was perfect—and then corrected himself to "all right."* So much for the flight. After half an hour I still think I can fly—and well. More reports as action develops.

The only thing I am at all worried about is Aerology, and I am sure I can get it. The fellows are a pretty good lot, not overly bright nor as fine fellows as those I left, but O.K. The atmosphere is not to be compared with Liberty. I find Lawrence a very nice little town, vaguely reminiscent of parts of Wichita when I was a wee tot, when the P.O. was at Market and (what is it, south of Douglas) and there were basement entrances to the 5¢ & 10¢ stores. Still hills, though, worse than M[issouri]. I am definitely going to enjoy my 3 months here and make them pay dividends. Went to the Christian Church this morning. Rather nice church, a little like ours. . . .

Had a nice visit and went to a show last night with Gailen and Jeanne W——. Both he and Charley are having finals next week and then get a two week vacation. Charley is going home, Gailen and his wife to Colorado for their honeymoon. Both are finding medics a very stiff grind and are a bit dubious of the final results. Charley is fed up with the army and doesn't look too well. None of the med students do, for that matter. All work and no play. Something the Navy doesn't believe in.

We are moving into our permanent quarters in a few hours, so I will close and get ready for another switch. Have been living out of my little hand-bag for the last three days, just got my greens out last night.


September 6, 1943


Dear Folks,

No news, just a plea. It seems I took a few too many trips to K.C. and Wichita on the last month's check and am tonight down to $1.03. That will neither see me through till the 15th nor see me home this weekend as I hope to do. Also I am low on underwear, socks, handkerchiefs, stationery, etc. So how about forwarding the laundry case plus the above-mentioned items and $10.00?

About the only news is that we are now in our permanent quarters.

---

*In fact, the instructor commented in Rea's Pilot Rating Book, "Fair student, somewhat tense. Good orientation. About average."

A fairly nice house, close to the campus, and not uphill. Not as nice as our first quarters, but an improvement over Wm. Jewell. We are three to a room—have a desk and most of a dresser for my own use. My roommates are the two other fellows from the Sixth [Battalion]. They are good roommates—quiet and mind their own business, which suits me. Too windy to fly today, so nothing else to report. . . .

September 7, 1943

Dear Phyl,

. . . You will note my stationery doesn't match and I am very tardy in writing, and am no longer in Liberty, Mo. To wit:—I have graduated a month early and am now in W.T.S. due to my being a superior intellect and fifth man in my Batt back at Wm. Jewell. It was really quite foolish of me, and I missed out on some valuable training. This is more than offset by the luck of being stationed at K.U., the best W.T.S. school in this territory. We live in ex-frat houses, eat off china instead of aluminum trays, live only 3 to a room, have lots of time to study and relax, lots of liberty, and incidentally, are at last flying. To date, I've had but an hour and a half, but will pile on the hours quickly. I seem to be doing very well, but it is definitely a job. Lots to remember and watch out for, and a quick-tempered instructor in the front seat. Mine is a pretty good old boy, cusses me out in the air and then encourages me on the ground. I take over completely for take-off, flight, and landing tomorrow and am a bit nervous about it, I'll admit. No great danger, but you can so easily make an utter ass of yourself. However, I shall do well. (Determination is supposed to be the worst half of flying).* Needless to say, this is heaven, Kansas, and close to home. Will be here about three months.

September 15, 1943

Dear Folks,

Not much news, but it's Wednesday night and taps doesn't come till 11:00 so I'm writing a few letters, starting here. The weather hasn't

---

* Of the September 8 flight, the instructor wrote: "Air slightly rough. Did very nice job. Has good coordination and good judgment. Very apt and absorbs instruction readily. Good student."

Communications class, WTS, University of Kansas, 1943

been so good lately. Got up yesterday in rough air, but neither Monday nor today. Had my first spins yesterday. Some fun. In a good tight spin, you go just about straight down and around at about one revolution every two seconds. The pull-out is swell, your stomach keeps on going a little ways and you can feel the pull in your face as well as in your "seat" (that's what you fly by). I'll admit I was pretty scared the first time, but the wings usually stay on, and as long as they do, you have nothing to worry about. We practice spins at 3000 feet and better, so there is plenty of time to pull out. One of the younger instructors took a fellow down from 4500' for seven spins, today, more than enough in a Cub.

Had our first code test today—4.0 for me. Have a 3.8 in Navigation, 3.3 in Recognition, 3.6 in C.A.R. [Civil Aeronautical Regulations], about a 3.6 or 3.7 in Aerology. All grades are on the rise, so I feel pretty good about it. If I could only get some flight time in soon. P.T. is plenty rough. Lots of good calisthenics and obstacle course. Tough. Longer than the one at Iowa Pre-flight by 100 yards. . . .

September 20, 1943

Dear Folks,

Here I sit, out at the airport, waiting for a couple of hours to fly, thanks to somebody's screwy idea we should spend more time out here. Spent a very nice weekend in Kansas City. Went in via the thumb route—trains and buses being late—and got in in time to attend the 6th Batt graduation banquet at the Phillips Hotel. It was a very swell affair, and it was great to see the old gang once more. Everybody was glad to see me and hear about W.T.S. As I said, it was a gala affair—all the officers praised the Sixth to the skies and said we were strictly tops. It was interesting—and a rare sight—to see Lt. Mason escorted down the main street of K.C. by twenty or thirty cheering, singing cadets—and everybody having the time of their lives. A very nice weekend, all told.
. . .

Haven't soloed yet—too much windy weather. Should be able to make it Tuesday or Wednesday. The laundry case came, thank you, and I'll make sure the cookies are well cared for. Please don't forget the gloves the next time. It's getting kind of cold up here. We're hoping to get some kind of jacket soon, before we freeze in the morning. My first week scholastic average was 3.4, but the next one will be better. Made 4.0 on first code test, last two recognition tests, and last week's Aerology test. 3.7 average in Navigation. . . .

September 22, 1943

Dear Phyl,

Wednesday night being liberty night, I hie myself down to the local cinema and see a show entitled "Claudia." It puts me into a very mild and mellow mood, so I come home and write a letter. Thought I'd wait until I had soloed, but the way things are going, it may be another week. One day it is too windy, then they change instructors on me, then he goes on a vacation, then it's too windy again. Right now it is raining.
. . .

It seems odd to walk around the K.U. campus and see the few civilian males still in circulation. They are a motley crew—shirt-tails out, pants rolled up, etc. Life on a big college (?) campus. The women

J3 Piper Cub flown in WTS

are an interesting and amusing lot. A bevy of comparatively young beauties were discussing prospective pledges while I sipped a malt this evening. Not much different from F.U., but they do get around more, *if* that is a help. I am beginning to realize that a person grows four years older in college as well as undergoing certain mental and psychological changes.

Golly! I don't know what's wrong with me, lately. At times I sound like I had a beard a foot long. Louánne M——'s last letter was addressed—"Dear Grandpa," the nerve!

September 23, 1943

Dear Folks,

Just a line after lunch to let you know that I have now soloed. Went up this morning and made four very pretty landings with no trouble whatsoever.* The fun of flying just begins when you get up by yourself and can really be master of the plane. My instructor was also quite pleased with the performance. Should have made it yesterday, but it

---

* From Pilot Rating Book: "First solo. Excellent job, uses good judgement."

was too windy. Today was smooth with overcast and wet field, but nice air.

Broke out my overcoat last night when I went downtown. It felt nice—and is a swell coat. We are still hoping for some kind of jacket or coat for regular wear with our greens. . . .

September 27, 1943

Dear Folks,

Monday morning and I'm standing security watch again. Spent a nice quiet, restful, cheap weekend here in Lawrence. Went to a show Saturday night, church Sunday morning. The other V-5 men have moved into the house next door, and we are all eating there now. Sunday dinner was delicious: breaded steaks, sweet potatoes, peas and carrots, some kind of baked peach, coffee, all the bisquits wanted with plenty of jelly, and ice-cream for dessert. And all very well cooked. Wrote some letters early in the afternoon and then spent a couple of hours in the K.U. museum with Gailen and Jeanne W——, who had just gotten back from a two week vacation-honeymoon in Colorado, down around Trinidad. I had seen the fine arts museum (Thayer, I think), but at that time—a fencing trip up here in 1940—the natural history museum was undergoing repairs. They have a very nice central room covering the animal life of North America pretty thoroughly. . . .

For Sunday evening chow we had sandwiches of lunch meat and cheese, potato salad and all the peach cobbler we could hold—and I mean *all*. I took twice as much as the house-mother thought I could eat, and then went back for seconds on everything. If I don't gain weight here it will be because they put in a 16 foot wall on the obstacle course, to be scaled with the help of a rope—that means climbing the rope and then switching to the wall. Am still getting 4.0 in code—now at 12 words per minute. Begin work on my Stage A flight sequence today—includes rectangular course, "S" turns, figure "8"s, stalls, and spins.

September 30, 1943

Dear Phyl,

Your letter was the best part of a day that was notably grey and dismal. It has been rainy for the last two days, which means no flying, which doesn't exactly suit me. I soloed a week ago and had the best hop I ever had. Subsequently I have been up and down but have certainly enjoyed my flying more than ever. I now fly alternate hours of solo and dual, which means there is a little time to practice maneuvers. This week has been a particularly bad one. On Monday, the fellow who flew before me smashed his tail wheel on landing and so grounded the plane for the rest of the day. Tuesday I was fortunate enough to get two flights. The first was a solo and just about as bad as I have ever had. You know how it is; some days you just can't get on the ball. I wasn't looking forward to the dual, but to my surprise (and the instructor's, also, I think) it was a peach.* Worked on my low stuff—rectangular courses and figure 8's at 600 feet, "S" turns and forced landings. It was good and certainly pepped me up. Then rain. It looks now as if we may fly tomorrow—with luck—but the field will be soft as mush, which isn't nice for landing and take-off.

Our long-sought-for Stearmans are at last appearing in the form of N-3-N's. The N-3-N is a similar model built by the Navy—not up to the Stearman, but a definite improvement over ye olde Piper Cub. I should get to fly one in a month—I hope.

October 5, 1943

Dear Folks,

Just a line before I start studying to let you know that all's well, but going a bit screwy. To start out with, I had a wonderful flight with Weatherly, the check pilot, on Monday. We practiced stalls and spins and he was very generous in his praise—"Alert, quick to catch on, absorbs instruction readily, etc."† Today I flew with Lassiter and the flight generally stunk (or stank)—however, a good forced landing helped.†† Then I pulled out a 3.8 in the big engines test that I thought I flunked, and at the same time slipped horribly on a navigation exam. Am still on the ball in Aerology and had better be back on in Code tomorrow. At present I'm nursing a beauty of a "charley-horse" in my left thigh—picked up in football yesterday. Could hardly walk today, but it's working out and is certainly good practice in self-control.

Everyone is worrying about their Stage A checks now. Mine is due

* Officially, "Progress Satisfactory."
† Precisely, "Alert—quick to catch on. Leveling off a bit high on landings."
†† "Weak on planning. Progress O.K."

Saturday, so I have three hours to practice and am confident of my work. I know I'll do my best and that will be good enough. Wind is the only ghost in the closet. It can play havoc with your low work. A number of the fellows have gotten down-checks and a few hours extension, and so far none have washed out. We're all hoping no one does. . . .

Oh yes, I almost forgot. Last night topped off the screwy happenings. I was supposed to have a watch from two 'til four or so; feeling tired, I turned in at 8:00. Woke up when everyone else came in at 10:00 and then rolled over for another four hours. Woke again at midnight when the watch was changed and settled down for the last two hours. Woke up again sometime later, but as no one called for me, went back to sleep as fast as I could. Finally woke up and didn't feel like going back to sleep, so decided to go downstairs and relieve the watch a little early. Much to my surprise, no one was on duty. The radio was going, so I sat down and listened until I got a 5:00 newscast. What had happened was pretty evident: the 10:00–12:00 watch had roused his relief who had mumbled something and promptly gone back to sleep. That left them both snoozing as 10–12:00 had promptly turned in. Luckily I was up at 5:00 and took the remaining hour's duty and rang reveille. Everyone was happy, as such things just aren't done, and we might easily have slept through until 7:00 or later. Old Johnny-on-the-spot Rea. That's me. We've been hearing a lot about how tough Preflight is, but from word I have received from fellows who are there, it isn't much different from Flight Prep at Wm. Jewell, just a little more advanced and four hours of P.T. So who knows, providing I pass the check Saturday—

October 11, 1943

Dear Folks,

Another weekend come and gone, and very quietly. As I told you, we all went to the K.U.-Iowa State football game Saturday afternoon. It was a pretty poor game, Iowa 13–K.U. 6. Iowa had a nice passing game, but K.U. didn't have anything. No one is quite sure just how the Jayhawkers scored. An all-Navy 440 relay was held at the half and won by the V-5 team. . . .

My check flight was much worse than I hoped it would be as I got very high marks all during the preceding week. However, they switched planes on me, the wind switched, and some of my work wasn't at all good, especially the landing. Weatherly told me I passed when we got down, but the way I felt about the flight, I'd rather have had a down-check and a chance to show him how well I can fly. Saturday, though, Lassiter told me he had given me an 87, which is plenty good, so I figure that if my worst is that good I shouldn't kick too much.* Began work on Stage B, Saturday. It includes all of Stage A, done with precision, and several new maneuvers. Our advanced coordination exercise is a beauty. You pull the nose up on a S heading and then let it fall off and turn to E, pulling the nose back up as you reach the heading, and then reverse the procedure. It's lots of fun and does take coordination. You have to lose and gain an equal amount in order to keep a steady altitude of 1000'. The "pylon 8" is a long slender "8" like this:

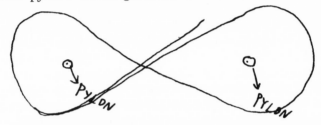

The pylons are pretty close together, so the bank is fairly steep, and still there has to be a period of straight and level flight between the turns. Takes careful judgement and a light steady hand on the stick. "Slips to a circle" are the most fun. A slip is just loss of altitude while maintaining a steady course with no gain or loss of flying speed. It's done by dropping

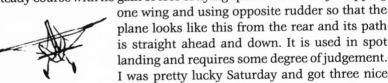

one wing and using opposite rudder so that the plane looks like this from the rear and its path is straight ahead and down. It is used in spot landing and requires some degree of judgement. I was pretty lucky Saturday and got three nice slips right into the circle on the auxiliary field. Lassiter was very happy and so was I.

October 12, 1943

Dear Folks,

It's a wet and windy morning—looks like we won't be flying for some time, so I'm getting my laundry off today. . . .

* The grade of 87 was generous; Rea's evaluation of the flight was accurate.

It was very interesting flying, yesterday. We had a wind of about 40–50 m.p.h. upstairs and you could fly almost straight up and you had to really dive it to lose altitude. Although it was a perfect day for a power-on landing, I had orders to make a regular 3-pointer, and it was ticklish business. Had to cut right in over the edge of the field, and by the time I got down on the ground, I had practically no ground-speed, just air-speed. Fun, though, and I made a nice landing.

October 19, 1943

Dear Folks,

Here I sit out at the airport waiting for my second flight this morning. The Intermediates having left for Del Monte [Pre-flight], we have lots of extra instructors and planes. Consequently, I am now with another instructor—a pretty good boy. Our primary trainers, N-3-N's came in yesterday. They are wonderful ships in comparison to the Cubs—real Yellow Perils, and are we hot to get on them. The instructors are just as anxious as we are to start using them. Flying two hours a day will mean that we'll be through Stage B faster than we expected. Finals for Elementary ground school are this week. Engines and Code at 14 wpm are going to be rough.

. . . Not much news. Keeping pretty busy. Very windy today, 50 mph at 2000' and a good 15 or 20 mph on the ground. May not fly any more today.

October 24, 1943

Dear Phyl,

Here it is Sunday afternoon, and I, enjoying the memory of a fine dinner, being soothed by the strains of a Strauss waltz on the record player, am standing, or rather sitting a watch for my roommate who is flying. At times one can find real enjoyment in even the Navy. . . . It's been a pretty full week. To start with we had time trials on the obstacle course Monday, and I figured I'd be better off running it every day than just once a week, like most people, so I promptly got on the sub-squad. At least I'll get used to Preflight practices, running all the time. Wednesday our final exams for Elementary began. I was feeling lousy and so was relaxed enough to get code at 14 wpm for a nice fat 4.0. Thursday was dedicated to more exams which turned out pretty well for me—4.0 in Navigation and other courses averaging 3.7, which is pretty good. Friday I had the CDO [Cadet Duty Officer] duty all day plus two hours early in the morning on Security watch. Main event of the day was being relieved to take care of the Lieutenant's one-year-old son while papa judged a K.U. beauty contest. Little Benny and I got along fine as soon as I discovered his love for poker chips! But then you never can tell about a child.

All this meant I had missed three days of flying—it rained Wednesday—and the weather looked bad for Saturday, so I was all set with tickets and reservations for Wichita and plans made for an enjoyable evening at home with the few fellows who are left. . . . So, in the middle of a downpour Saturday morning we were informed that we would fly Sunday. This was slightly upsetting as the C.O. had said previously—no Sunday flight unless announced for the following week. So, a telephone call home and cancel all plans. I am in a foul mood, and a flight principles test doesn't help. Half the fellows flunk it, and I am just mad enough to choose the right answers and get a 3.5. The remainder of the afternoon we have liberty, which I spend trying to chase down a laundry bag and finally being stuck $1.25 for a $.25 article. So I go into the music store and listen to a Liszt concerto vying with Boogie Woogie in the next booth. At least I thoroughly enjoyed Monty Wooley in "Holy Matrimony" last evening. A most delightful character.

Some two and half hours passed since I started this letter. It is now one of those sharp October afternoons when to sit in a soft easy chair listening to the symphony and watching the leaves shiver and fall outside is to live again a thousand such afternoons, perhaps a thousand lives; the shadows of the past become more real than the reality of today. It is a state of mind I greatly enjoy at times like this—a state in which individuals and incidents may be introduced, or excused, so easily.

October 25, 1943

Dear Folks,

I'm standing the graveyard watch again tonight in order to be free next weekend. I was quite disappointed that our plans for the past weekend didn't work out, but it was one of those things that can't be helped. Tuesday and Wednesday the weather was bad, Thursday we had finals, Friday I had C.D.O. duty, so according to the chief pilot there was time to be made up. Friday Night Lt. N—— said there would definitely be no flight Sunday, but Saturday noon he was forced to announce the flight. Everyone was pretty disgusted. We had had no liberty all week and some pretty stiff exams to boot. . . .

The tests Wednesday, Thursday, and Saturday were plenty rough. I was feeling lousy when we took the Code test and was afraid I'd only get 10 wpm, however feeling bad took my mind off code so that I relaxed and got 14 wpm easily. I was happy to see another 4.0 in the Navigation problem which was a beauty. Engines, which had me worried, came out 3.4; C.A.R. 3.7, Aerology 3.6, Recognition 3.6 for a 3.7 average, to be topped off by a 3.5 flight test which the majority flunked, or nearly so. With a normal amount of luck on my flight check this week, I'll complete Elementary with a very satisfactory record, and one of the best in this group, if not the best, which is possible.

Despite everything, yesterday morning was wonderful flying. We had a pretty good wind and lots of low clouds. On the first flight the ceiling was down around 800 feet at times and it was no trick to get up into the clouds and lose all sight of land. I went up through it to 1500 [feet] and decided I'd better not get up too high, and it extended up to 2500. Visibility under such conditions is absolutely zero and can change from unlimited to zero in a hundred feet. The second period, about 10:00, the sky cleared and I went up to 3000' to practice spins, only to have the clouds close in under me. The clouds look beautiful from the air and extended for miles yesterday, but I like to keep one eye on the airport. Also made my first solo power wheel landings today, or rather yesterday. The last one was a beauty, so smooth you could scarcely tell when you were down. The procedure is just the opposite of a normal power-off 3-point landing in which the plane is stalled in order to lose flying speed. In the power landing the plane is flown onto the ground and the tail raised in order to decrease the lift on the wings. Then, when speed is lost, the tail settles gently to the ground. It is a case where theory works beautifully and is much easier and safer than it sounds.

October 26, 1943

Dear Folks,

Well, I'm just about through with Elementary. Tomorrow I should have my Stage B check. Am not worried about it, but hoping for good luck. Final results of the government exams showed Cadet Rea as high man in the 44C class. It was announced today that 16 men are leaving Tuesday for Iowa Preflight. The top men were given their choice of going or staying a possible three more weeks. This one chose to remain for several reasons—not the least of which is that I enjoy the proximity to home. Also I want some time in the N-3-N's which we should start flying Monday.

Intermediate should be easy enough—no Engines or Recognition, just Code, Navigation, and Aerology.

I finally got permission to attend the concert last evening, but had to spend some time in an Engines make-up class, having missed an hour while on duty. Therefore I arrived just in time to hear the last high note of the Mad Scene from *Lucia*. The remainder of the concert was very wonderful. . . .

Will be home on the Streamliner Saturday night and would like it if last week's plans could be carried out. The cake is swell and going fast. The laundry case is on its way, and I'm out of paper.

October 28, 1943

Dear Folks,

Just a line before lunch to let you know that all is going well with Cadet Rea. Had my Stage B check today. Flew with Chief Pilot Miller, a tough check, and got a 90 on it. He had no comments for me and told my instructor, Robertson, that it was one of the best flights he had ever had. Even my slips were so good he said he would hate to have to demonstrate them to me. It was a swell flight and not under the best conditions, either.* Really makes me feel good. Now to get started on the N-3-N's—Monday, I hope, although there is some talk that we may fly Cruisers for a while—they're just a deluxe model Cub with a bigger engine. I'm not sure just who will be my instructor in Intermediate. May go back to Lassiter or I may have Weatherly. Robertson is taking on a group of beginning students who came in Tuesday.

Last evening we took part in a big Navy Day parade in which all the K.U. military organizations participated. The whole town turned out to see the parade and, I hope, got as big a kick out of it as we did a big march. . . .

* The Chief Pilot judged "Student far above average."

October 29, 1943

Dear Phyl,

. . . My work is hitting a new high at the present time. Last week we had finals for Elementary and who should show the boys how, but Cadet Rea. Top man in the company for grades, so why not in flight also. O.K., so I go up yesterday morning and give the chief pilot one of the best rides he ever had, so he reciprocates and gives me a 90, which is also high. Everybody was very happy about it, my instructor as much as myself. You would have thought someone had a baby the way he and my Stage A instructor came around shaking my hand and saying nice things. So Monday I start in on the Intermediate work in the N-3-N's. I really hate to see the days go by, Phyl, I'm enjoying flying as I've never enjoyed anything before. It gets in your blood in a way that's almost incredible. When you make a perfect three-point landing at the end of a good flight you're really on top of the world. Whether it's for the Navy or not, this boy is going to fly for the rest of his life, believe me.

Sixteen men from our group are going to Iowa Preflight next week. The top men were given their choice of leaving or remaining for two or three weeks of Intermediate. I chose the latter course for several reasons—flying, proximity to home, and I don't particularly want to go to Iowa. Previous groups have gone to Del Monte, Calif., and I'm willing to wait and chance it. Feel a bit guilty about not going ahead when I have the chance, but figure the time in the heavy planes will more than make up for two weeks lost now when I get to Primary.

October 31, 1943

Dear Phyl,

I am now in Intermediate. Passed my Stage B check with a good high 90 (did I tell you this once before?) and so left Elementary at the head of my class. Start the N-3-N's tomorrow morning. I don't know when I've been happier with my work than at present. It's really a swell feeling to be in love with your work and be doing well with it at the same time. I just hope it will continue to be that way.

Got home again this weekend. Had B——, T——, and O—— over Sat. night, and we just sat around talking, eating, and listening to some music. . . . Nice, but makes it unpleasant to leave and return to the job at hand. It's a funny feeling I get. Not homesickness at all, something different. Instead of wanting to go back to the past, I find myself wanting to go into the future. . . . I often wonder what—and when—that reality will come to pass. How will it be? And again, will it be? It's a long road, and yet when I think that I have lived, say, a third of my life—accomplished so little, I feel I dare not wait an extra day. The ever-present possibility that one's days are numbered. Very silly. Time is something man has a sufficiency of, whether it be years or seconds, it makes no difference. Some people are being hit hard by the war. I'm gaining thereby and feel a bit guilty about it. I hope I can make it up to them sometime. . . . I'd be a liar to claim any desire to be a hero, but I'd feel a darn sight better if I were doing something to get the thing over with. Ha! I'll probably have my fill of that before too long.

November 1, 1943

Dear Folks,

Well, today I took my first hop in an N-3-N. It's quite different from a Cub; in fact, there's very little similarity outside of principles of flight. First there's the open cockpit which is cooler but allows real vision. The controls are much more sensitive and larger (pedals, stick, etc.). The instruments and controls all work smoothly and efficiently. My only trouble was that I have to reach for the stick; however a few cushions should fix that.

The take-off is quite nice; plenty of power to get you off at about 50 K [Knots: nautical miles per hour]. She climbs at 65 K and doesn't waste any time going up. Cruising speed is about 80 K without any wind. In comparison to a Cub, the N flies much smoother and handles nicer, reacting to the slightest touch of the controls. You might compare the two like the Ford and Cadillac cars. However the size and build of the N reminds me of Dad's Diamond T. The plane lands nicely, tail first for the Navy, and a smooth landing isn't necessarily a perfect one. I have a great deal of respect for the N. There's a lot of plane there, and you've really got to fly it. Did a couple of spins, and the recovery is somewhat different from the Cub and not as easy to bring out on the heading. But she stalls and spins beautifully. It will really be O.K. when I get used to it. Should be able to solo after about five hours of dual time. . . .

N3N-3 Navy trainer, cadet (not Rea), and Chief Pilot Miller, WTS, Lawrence

November 7, 1943

Dear Phyl,

I'm in a rare mood tonight, Phyl. It started Saturday afternoon when a select group of six helped with the funeral services of an ex-aviation cadet who cracked up at Corpus Christi early this week. It's a bit sobering to carry the casket of a fellow cadet in whose footsteps you are following and realize that sooner or later your luck may change, just as his did. There was a definite finality that was almost coldly cruel when his buddy, who accompanied the body up from Corpus, handed the flag which had covered his body to his mother, turned, saluted the no longer empty grave, and taps was sounded.

So in this delightful frame of mind, and with a good strong cup of coffee under my belt, I went home and spent an utterly wasted weekend. I'm going to be quite content to be far enough from Wichita that return is out of the question, sometime in the next few weeks. My friends, my family, the old places, all mean a great deal, but I no longer wish for the old times and places. I want a completely new existence with a new selection of interests, purposes, ideas that will be new only in that I will live, rather than just dream them. . . .

Flight goes well enough. I'm really getting a kick out of the N-3-N's we're in at present, and they're plenty of work to fly well. Still no word of when or where we go. Probably in three weeks at most, maybe less. Celebrated one year in the Navy, Nov. 4.

November 8, 1943

Dear Folks,

It's a pretty cold, grey day here in Lawrence. We're wearing our sweaters, jackets, and overcoats with lining for warmth between buildings, topped by watch caps which are black wool stocking caps. No flight this morning because of snow flurries which have developed into a steady descent of fine dry snow. The boys from California and Louisiana are having a great time. At least half a dozen had never seen snow fall before last night, and most had never seen anything comparable to today's little blizzard. It's nice to look out at, but the rest of us would just as soon it came to a halt.

We had a special ground school session in navigation today. Got a test which was the worst I've ever seen. They seldom get me down, but this one did. With twenty minutes left I finally wrote my first answer. I failed to finish by my final arrival position, but salvaged a few points from the wreckage. That's the first time I have failed to finish, and the last, I hope, although two minutes more would have done the job. . . .

November 10, 1943

Dear Folks,

Here it is a perfect day for flying, and I'm grounded, thanks to a broken starter. Yesterday we flew in weather I'd ordinarily not even go out in. Boy, it was cold! Had on my black wool sweater, windproof jacket, lined flying suit, wool socks, overshoes, helmet, goggles and gloves, and nearly froze my fingers and face. We all went up solo yesterday and the instructors sat around drinking hot coffee and congratulating each other on their cleverness.

We've lost two men this week on flight washouts, and another fellow is dangerously near Board action on disciplinary reports. Monday we had a Navigation problem that was a honey: something like this—leave ship CV for ship FG at 0800. Upon reaching FG, go out on a geographic sector search from 250°–220°, returning to CV at 1100. Sight enemy during search and follow him. Before you are ordered to return to ship, enemy changes course, and so does CV. You are finally ordered to return to your ship after checking the wind, running a wind star, and assuming a new speed. Finals next Wednesday. Aerology is quite interesting now. Flight planning with actual weather conditions. Takes some thought and analysis but is certainly worthwhile. . . .

November 16, 1943

Dear Folks,

Not much news, just a last line from Lawrence. Have spent an easy two days winding things up here. Finished flying with a total of fifty-one hours—fifteen in the N-3-N's. Have packed a few things and am not sure just how much more I can get into the old laundry case and bag. Guess I'll put my dirty clothes, shoes, and gym gear in a heavy laundry bag, tie it up well, and ship it that way. I understand that we have to purchase new luggage at Preflight, so I should be well supplied by the next time I move. Heard from my friend W——, who washed out and is at Great Lakes; he says it is a sort of rest cure for tired cadets. No assigned work, and liberty every other night. Lots of cadets from Corpus and Pensacola who were about to get their wings. The whole bunch here is pretty serious. No one particularly wants to leave, to quit flying for three months, or go to Iowa. The grind that made us glad to leave Liberty will begin all over, and washouts will come faster. But that's the way it goes, and if a fellow fails to get the breaks, why he's in the Navy and there's still a pretty big job to be done, even if not in the air. . . . We are leaving at 2:40 tomorrow afternoon and will probably reach Iowa Thursday morning. I'll write as soon as we get settled.

November 16, 1943

Dear Phyl,

So this is the last letter from Lawrence. Tomorrow afternoon we are on our way, ten of us, to Preflight at Iowa City, Iowa. Our dreams of

N3N-3 Navy trainer flown in WTS Intermediate training (Courtesy U.S. Naval
Institute, Library and Photographic Services, Washington, D.C.)

California dissolved as do so many. I have been exceedingly fortunate
in my training up to the present time; I trust it will continue. Heard
from a buddy who washed out and is now at Great Lakes. He says that
it is a rest haven for tired cadets, some of whom have been there since
April. So maybe it won't be so bad if and when.

We've finished off our work here with over fifty hours flight time,
and fifteen of that in the N-3-N's. The last couple of hours were spent
just messing around, dog-fighting, spinning, and having a good time in
general. It has been an awful lot of fun, these past two and a half
months—now we get the works. . . .

Had liberty tonight, so I went downtown and saw "Watch on the
Rhine." That, and leaving and everything, makes me kind of morose,
or at least blue, tonight. Ten percent of our group washed out here; the
percentage may easily be higher in Preflight. It's not so much a disgrace
to be eliminated—it may well be for your own good—it's the idea of
starting something you can't finish. So, like a bunch of crazy kids, which
I guess we are, we'll probably spend our stopover in K.C. getting drunk
and spending the month's pay check.

# 4

# Winter of War

I N A L E T T E R T O Phyllis Edwards on January 22, 1944, Bob Rea spoke of "fog and clouds and the turbulence all around which is our life today." It was a description that the people of Berlin could have appreciated. On the night of November 18, 1943, when Rea had just arrived in Iowa City, Iowa, to begin the next phase of his naval training, the RAF commenced "The Battle of Berlin," a campaign whose twofold objective was to smash one of the major industrial centers of the Reich and to shatter the morale of the inhabitants of the German capital. Three night attacks within a week brought a more brutal reality to that population than it had ever before experienced. Yet, despite the pall of smoke, the wreckage of buildings, and the grief for 10,000 people killed in the three raids, Berlin was "neither destroyed nor had its people been shaken into panic or revolt."[1]

In Italy, Allied casualties were being carried on the backs of men and mules down the muddy slopes of cruel mountains. Life and death were close companions that Christmas. Air crews aboard the fast carriers in the Pacific were more comfortable and might celebrate the holiday in some style if they were not engaged in strikes against the Marshalls, the Navy's next objective. It was, however, no more pleasant to die in water than in mud.

At home, hints of happier prospects were to be found in the first issue of *Life* for the new year 1944: "Washington is now busy with the semisecret plans for the reconversion of some portion of industry to civilian production. Bobby pins are on their way back, and present schedules call for a 1944 production of 2,000,000 electric irons, together with a certain number of hot-water heaters, baby buggies, etc." *Life* also reported that the first surplus Jeeps had been sold, at a price of $750, to lucky American civilians.[2]

Bob Rea embarked upon another phase of training during the winter of 1943. It did not take him into the sky—fortunately, perhaps, for it was a season of mists. Pre-flight was a name to conjure with, for Naval Aviation training was characterized by its Pre-flight Schools. In them, boys were transformed into men; the weak fell by the wayside, and only the strong survived. Pre-flight was symbolized by the dread Obstacle Course—that ingenious means of testing the agility, strength, stamina, and determination of the nation's youth and guaranteeing that only the deserving won those wings of gold. The school located at the State University of Iowa had a forbidding reputation for severity which the Navy did not discourage.

Cadets were housed in university dormitories, but the Navy was much in evidence, both the old Navy, epitomized by Captain D. C. Hanrahan, the cadet corps' hero, and the new Navy, embodied in the young ensign from Alabama who mustered Platoon F-3, Thirty-ninth Battalion, outside the barracks in poplin windbreakers when the temperature at 0600 was ten degrees above zero. The prevailing impression of Iowa City Pre-flight was cold, cold, cold—gray and cold.

Knowledge was advanced in the now usual subjects—slowly and not very painfully. Celestial navigation was fascinating to Cadet Rea, and his comments on naval history are amusing in retrospect, for in later years he would write some naval history himself.[3] Physical Training dominated the scene at Pre-flight, just as the Iowa Field House dominated the campus. Presumably it did some good. Rea was less than candid in his correspondence regarding his experience in the competitive sport of boxing. It was, in fact, a great relief when he was put on the subswimming squad for failure to glide through the water in the official Navy Way. The pool was ice-cold, and the episode of Captain Hanrahan's visit was truly memorable. Strangely, the famous obstacle course seems never to have raised its ugly head.

Veterans of those years will remember the long waits in telephone queues and the longer delays in getting calls through. They were a part of life during World War II, along with food rationing, gas rationing, radios, and phonograph records. While Cadet Rea was at Iowa Pre-flight, Phyllis Edwards began teaching at Westbrook Junior College, Portland,

Maine. Perhaps a winter in Maine helped to warm her heart; the letters she received from Iowa City were not less lengthy or frequent, but they became less concerned with naval affairs.

In spite of the Tree Squad and temporary deafness in one ear, Cadet Rea made an early escape from Iowa Pre-flight, happy to be heading for Kansas and anticipating that the Navy would now get on with the business of flying.

# Pre-flight School, State University of Iowa, Iowa City, Iowa, November 18, 1943–January 31, 1944

November 18, 1943

Dear Folks,

Well, here we are at last—Preflight. Haven't seen much yet except cadets and ensigns. Chow is in a small nice room, all you can eat, and pretty good. Our room is plain; nice linoleum tile floor, three windows, first floor. Have already met a number of my old buddies—none of whom have washed out. There are four in the room, Bob P—— from Louisiana, Randy R—— from Arizona, Ed K—— from California.

(Next morning) From the looks of things, there won't be too much time to write for the first week or so. Just finished breakfast and don't know how long I'll have to write, so just details. Announcements are made over a P.A. system. Reveille sounds at 0530, chow shortly after. Measured for dress blues yesterday and should have them in time for liberty two weeks from Saturday—none till then. Off from 3:30 p.m.– 10:30 p.m. Saturday, 1:00–6:00 p.m. Sunday. Church attendance compulsory. I've signed up for choir and was drafted for station paper. Nobody here seems to think the place particularly tough, if anything, just dull. Like N.F.P.S. Strictly military. Commissioned officers for platoon officers. Ensigns and [Lieutenants] J.G.'s all over the place. Uniform of the day is green pants, brown shirt, black tie, green cap, brown G.I. shoes, and a blue jacket (slip-over) we have yet to get. Have to send home practically all civilian stuff, so my bags will be full of pajamas and stuff. Can't have any here, so no laundry case. Send my shirts, skivvies, and any other G.I. stuff, as soon as possible. Wrap in heavy brown paper would be best, I think, as I have no place for a box. We will buy new luggage when we leave here, and $30 a month is deducted to have the cash then. That means pay is $25 a month for practical purposes. But there's no place for to spend it, so—. Mail call three times a day, they tell me, so how about some mail. Will write when I have the time and know something more.

Am not yet impressed with the state of Iowa, although I slept all the way from Des Moines to Iowa City. Never saw so much fog and haze. Saw the state capitol, and was duly if not greatly impressed. Just rolling country and lots of corn. Our Platoon Officer formerly was quarterback at Alabama U.

November 21, 1943

Dear Folks,

Well, here's the lowdown. We came into Iowa City late in the afternoon after missing a train in Des Moines on purpose. Saw the town— not at all bad, but lots of smoke and fog. Went out to the old capitol after we saw the town. An old building in an odd style.

The scenery between Des Moines and Iowa City was just like eastern Kansas, so I slept into town. We were met and escorted up to the campus, logged in, and measured for our dress blues and whites. The next couple of days were so busy I can't keep them straight. We have had books issued, hats and covers, green coats, gym gear, blue windbreakers which are about all we wear outside. Bought a pair of gloves, too. Took a navigation test and made a 3.9. Not too bad, but should have been a 4.0.

Have had a lot of physical tests. Twenty foot underwater swim— passed. Two hundred yard stroke test—got jerked on breast stroke, but so did about half the fellows. The pool is beautiful—100 yards long and ten or twelve lanes wide, located in one small wing of the field house which is beyond description. Passed the physical fitness tests O.K. On the pack test (five minutes of stepping up and down on a bench with 50 lb. pack) I got a score of 93 which is very good.

We had two inspections Sat. and today; our room really shone. Chow is consistently good. Nearly half a chicken this noon, all the milk you want. Church services this morning were quite nice. The station choir of nearly 100 men is about the best male choir you ever heard. The entire station holds about 1800 cadets and probably 200 officers. Gold braid all over the place. Ensigns don't even rate. Passed the same physical we had at K.C. examining board in good shape. If anything, my eyes are in better condition. Got shot for yellow fever. One man and about five assistants handled us at the rate of one cadet every two seconds. You didn't even stop walking. That's no exaggeration! Even the fellows from the fleet swore they'd never seen anything like it. This place has a degree of efficiency that is amazing.

Reveille is at 0530. Before the second note has sounded there's an officer in the room yelling, "Hit the deck!" and we're already in mid-air. Twenty minutes later you have shaved (?), made your bunk (which you have to completely strip for 15 minutes), cleaned your room, and mustered for chow. Saturday we were in navigation class at 0645! When the P.A. system sounds off "Ten minutes until a certain formation," you pick up your cap and start walking for the door. If they say five minutes you grab it and run; if they say "immediately," that's tough, you're already late! It is quite cool, and we are only wearing the blue "breakers" which aren't as heavy as a G.I. shirt. Boy, we freeze. It's a penetrating type of cold.

Iowa City is across the river from our barracks. I don't know when I'll see it. Liberty is from 1530–2230 Sat. night, and 1300–1800 Sunday. We are restricted for two weeks to barracks, and all sub-classes come during those hours. This field house I mentioned: it's about as big as four city blocks and five stories high. In the main arena, which is used for basketball, football, track, movies, drilling, and Divine Services, you could lose anything comparable in Wichita. The Army and Navy work out at different ends simultaneously, and you can hardly hear the other group—no kidding. . . .

This has been an erratic letter, but we have a formation shortly to hear a Marine pilot. Our regular schedule starts tomorrow. I don't anticipate any trouble with anything. It's very much like Flight Prep only more advanced. We are due to graduate about Feb. 6. If grades are high enough and all else satisfactory, there's a chance of going out two weeks earlier. No time off Thanksgiving—just turkey. Should get the day off Christmas. The scuttlebutt has it that we'll get four days leave between here and Primary. If it continues to get cold, I think I'll be in the mood to put in for primary at Dallas. Some of the fellows want Minneapolis, but I've had enough cold flying to know when I'm well off. . . .

November 23, 1943

Dear Folks,

Well, time out for a few lines. So this is Preflight! It's now 2010. I got up at 0600 and left the mess hall at 1955. Our schedule is something like this—first class about 0725 and on until 1220. They seem to vary every day, but include code, blinker, naval tradition and etiquette, recognition, navigation (celestial), military (gunnery), and aerology. We haven't really gotten down to work yet, but none of it sounds too tough. I've got a pretty good foundation for all of it.

When I say the chow is good, I'm not merely refraining from griping. At noon today we had a *very large* steak, potatoes and gravy, creamed carrots, lettuce and mayonnaise, rolls and butter, cookies and chocolate ice cream, and all the milk you want. Tonight I had four glasses. There are two mess halls and both are very nicely furnished. We're still using the old navy trays, but that detracts little from the chow.

The afternoon begins at 1345. Our first sport is gymnastics—tumbling and apparatus. Second sport is basketball. Rest period from 1530–1600. Boxing from 1600–1720. Swimming, 1750–1830—this is the subclass which numbers about 150. Most of us are there, not because we can't swim, but because we don't do it the Navy way. And it is a nice, relaxing way to end the day, I think. Boxing hasn't developed much yet. We had an elimination, 1-round, and I was, to be honest, soundly trounced. However, you only learn by doing. I just hope I can learn a little before I have to go out for blood. Our first two sports change after two weeks, and I'll be in boxing for six weeks, plus two instructional. While I have to laugh at my own efforts to box, I can honestly say there are plenty of the fellows who are no better, and we have a good time kidding each other.

The evenings are pretty well filled up, as we finish chow just in time to return to the barracks and have study hour from 2000–2100. Taps at 2130 comes none too early. With good food, *plenty* of exercise and sleep, and good *cold* air, everyone is putting on weight. I weighed in at 136, which is the most I ever weighed. All in all, I'm enjoying it and feeling better than ever—even if plenty tired by taps. Only one real gripe—the bunk is lousy. I've sworn to stay with this outfit until they give me a really good bunk. On the hard one I have now, all I do is dream about flying. Oh, well, Primary is only 10.5 weeks off. . . .

Thanksgiving Day
November 25, 1943

Dear Folks,

This has been a pretty nice Turkey Day, considering. We were secured from 1530 on, and I had good company and the package of cookies and candy to while away the few hours of liberty. The week has gone very well, and I haven't a single gripe.

We had turkey dinner that was all one could ask for and more than we had time to eat. Absolutely everything—even pumpkin pie and nuts! Only the drinking water here is lousy—a sort of chlorine cocktail, but if it is cold enough, you don't notice it.

Now that I have all of my clothes, I'm well supplied. These cold mornings are pretty brisk, and the southern fellows don't care much for it. However, it warms up considerably by afternoon. Haven't received our blues yet, but may soon.

Studies are coming along fine. Recognition, which has been one of my weaker subjects, is really taught here. We are going to concentrate on the American and British fleets. Celestial navigation is very interesting and so far easy. The teacher is a [Lieutenant] J.G. USNR, former physics prof on the west coast and really knows his stuff. Graduate of the Miami navigators' school and can explain things so that they are perfectly lucid. And a swell fellow to boot. I think we'll start on the rifle range tomorrow. Been having a number of related lectures by the station's Marines. Code and blinker won't be too bad, although I am rusty on the latter. Can take code at 10 wpm here, which is as fast as 14 wpm at W.T.S., and blinker at 6 wpm, which is good enough to get out with.
. . .

November 28, 1943

Dear Folks,

One down—ten to go. Got weekend liberty this week and saw Iowa City. A nice little town—clean, plenty of stores, and quite conveniently located. Church services today were very nice. Attended choir practice, but won't sing with them until I get my blues, which should be this week. . . .

Saturday afternoon we went down to the rifle range and fired ten rounds. They don't care where you hit, just so all the shots are close together. Shooting breach-loading, single shot, 22 longs at 50 foot range, prone with a sling and peep sight. Nice easy shooting. I knocked out the bull's-eye with three shots and scalloped the edges with the rest, firing as fast as possible. The Marines run it and have some pretty elaborate training equipment. And boy, are they proud! Marines, you know, only about a dozen among 2000 Navy men. . . .

November 28, 1943

Dear Phyl,

Your letter this week was most welcome. I've been busy enough up here to keep from missing mail—but it had been a darned long time. I'm awfully sorry to hear why you hadn't written and hope that by the time this reaches you, you'll be up and at 'em stronger than ever. Maybe you should take a hint from the Naval Aviation program and take a course in relaxation. We are actually to be given such a course, and every man here is, I believe, an expert at relaxing when he gets a chance. However, we don't need to relax for any mental strain here and, in my opinion, don't need to relax from any physical exertion, either.

It's a funny place, Phyl. Most of our officers are from the dry land Navy, but seriously impressed with their job. Our platoon officer is a former football star from Alabama who is a perfect physical specimen and beauty, but without brains enough to drill his platoon or explain an order. It's like a game with a bunch of overgrown infants playing at being soldiers. We study gunnery, bayonet, naval reg[ulation]s and etiquette, and it's all sort of simple. The physical training is good, no doubt, but everyone feels that we're just marking time for 10 more weeks until Primary. But perhaps I'm over-pessimistic because of having to stand inspection for an hour this morning.

We're in the middle of our training now, and the fellows are beginning to realize that some seven months have gone by and won't be regained. Some of them are beginning to look like men instead of boys. They dress in their blues and go to town for liberty, but the gayety and fun of a few months ago seems lacking. You don't (and I don't know what came next. Church being compulsory, it is some seven hours since I began to write).

December 5, 1943

Dear Folks,

Methinks it's been some little time since I wrote. No excuses except I've been about as busy as possible. The week has gone slowly in ways, and yet busily enough that I don't notice any drag.

Academics are coming very nicely. We had our first test in code and blinker, and I did quite well, much to my surprise, as it had me worried. Celestial is coming along fine. All my grades up to Saturday were 4.0's, and then I transposed a 279° to 297° when I plotted a fix and so have a very ordinary 3.0 to keep my spirits on an even keel. Spent an hour early in the week studying stars used to obtain fixes. We are supposed to be familiar with about 22 before we leave. Some 65 are used in navigating. The work itself is at present quite simple and is largely comprised of looking under the right number in the right table in a book. We've had lectures about every day on guns and gunnery by the local Marine Corps. They are about as bored giving them as we are hearing them, I believe. We spend one hour a week on the range, firing 10–20 rounds.

In E.N.S. [Essentials of Naval Service] we are learning something of the internal functionings of the naval service. But slowly. Things like leave, pay rates, travel allotments, naval courts. Recognition is interesting, if a bit rough. We've gone through the U.S. DD's (destroyers) and all CA's (heavy cruisers) except the USS *Wichita*. We're starting on German planes next week.

P.T. has been nil except for one boxing match Tuesday, which netted me a stiff headache and sore nose for a couple of days. We're having labor engineering, which means doing all the odd jobs around the station's many offices. Soft duty! Next week we have football and hand-to-hand [combat]. Am still on sub-swimming, but by the time I get a chance to pass the tests, I'll be well able to.

That takes care of routine stuff, I guess. Am officially in the choir now. This Sunday we sang "Lo, a Voice to Heaven Sounding," by Bortniansky. Very nice. Next week we're doing F. Melius Christiansen's arrangement of "Fairest Lord Jesus." The only trouble with singing in the choir is that you can't hear the chaplain. Incidentally, our Chaplain was in the Coral Sea fight aboard the USS *Vincennes* which was sunk.

Wednesday night we all went to the show in the Field House for all station hands. Free, and a nice break in the week. Big event in the week, of course, was our getting our uniforms—two sets of whites and one of blues. Boy, were we the happy boys! Everybody tried everything on and ran around in all conceivable stages of dress and undress. Mine fit very nicely, and I'm quite pleased with them. Made our first liberty in them Saturday night and gave Iowa City a thrill—along with the other 2000 cadets. Everybody had pictures taken yesterday. We went to the largest studio in town, and the line was 24 men long when we left. All the rest in town were just as full. However the finished pictures should be in my hands in a couple of weeks, so I'll try to send you one for Christmas, providing they turn out well.

Speaking of Christmas, we're having a very nice deal here. Christmas Eve will be a big Regimental Ball—taps at 2400, liberty from 1530 on. Liberty on Christmas day following morning chow until 2230. And the next day is Sunday, to boot. Of course some of the fellows were hopefully listening to the three day leave scuttlebutt, but this is more than I expected. . . .

One thing about having a uniform and wearing a white shirt—you feel like a white man. And look like one. These navy blue serges and caps with the gold braid and stuff make you feel as dressed up as a civilian wearing a tuxedo.

Next week begins our winter schedule here at Iowa Preflight. The big change, besides P.T. in the morning and classes in the afternoon, will be that the first and only reveille will sound at 0630 instead of 0530. Quite an improvement. Taps at the usual time, no activities after 1720 or 1750. This adds an hour of leisure time before evening chow and gives us an hour or 30 minutes more sack time. We've really been lucky so far. The next 40th Batt, made up largely of 6th Batt men [at N.F.P.S.]

came in Wednesday. Several of the fellows from Liberty are here and getting a good ride from us old-timers as "boots."

Just back from dinner. Half a chicken, mashed potatoes, green peas, cranberry sauce and pumpkin pie. Makes up for several not so good meals during the week. Yesterday we took a hike along the river road. Really beautiful countryside we saw. A little like Kansas and Indiana—sort of in between, I guess. Distinctly Iowa though. Lots of ponds around the campus. Quite a campus, too, scattered over a good number of miles. Every type of architecture, but pretty well sticking to old English crossed with modern American. Beautiful landscaping, though. See a lot of grey squirrels around here, too. . . .

December 5, 1943

Dear Phyl,

. . . Two weeks of Preflight and I'm still whole and hearty. Of course next week we have football and hand-to-hand, so I may not feel so good about it next week. Had a most amusing round in boxing last Tuesday in which little Bobby conclusively proved he is no pugilist. Having been deposited none too gently on parts unmentionable three times in one none too short minute, the coach took me by the shoulder and in a fatherly tone said, "That's enough, son." I heartily agreed, but forbore telling him. Boy, how my head ached the next day! . . .

Got our blues this week and they really look swell. Spent a big liberty having pictures taken, getting hair cuts, and seeing a show. That's the way it goes. Get a good uniform and there are two thousand fellows with the same thing.

December 6, 1943

Dear Folks,

In order to keep up with myself this week and not get as far behind as I did last week, I'll get on the ball right now. This is the first day on the winter schedule, and I think it is going to work out very well. Reveille at 0630 is wonderful. That nine hours of sack time really goes swell.

Chow at 0720, so there's plenty of time to clean up and get the room in order. Muster for sports at 0815, so we can leisurely come back and change to gym gear. P.T. is plenty strenuous. Mass exercises—calisthenics, Navy style—football, hand-to-hand, and swimming, now an hour instead of forty minutes. Out at 1125, chow at 1200, again plenty of time to get from the field house to the barracks. At 1245 we have military, followed by academics. The day is over 1730, which is a lot better than 1850. Both morning and afternoon pass much more quickly. There are rumors that we'll be off at noon Saturday, too.

The best one came today when we were issued bayonets! And to top it off, they are made of plastic instead of steel. They look nice but don't seem to fit into an aviation program. Thought we were due for some snow today. It rained all night and was cold and cloudy this morning. Cleared up, though, and was a very nice afternoon for our hike to the armory. . . .

December 9, 1943

Dear Folks,

Well, I just completed my first day of watch at Preflight—24 hours of it, although not continuous. In view of the shape in which the fellows returned from P.T., though, I'm not sorry to have missed the day.

It has really gotten cold today, biting cold, and little heat in the buildings. Working in the Batt office, I kept tab on all the guys in sick bay and the hospital. Twenty men in the latter and nearly half of the rest went to sick bay for one ailment or another—mostly colds and coughs. Almost everyone has a cold, so you get plenty of sympathy.

Not a great deal of news this time. We got our first grades in E.N.S. and Gunnery. The E.N.S. test was a corker. None of the right answers were right, and fully one third of the Batt flunked it. The average was about 2.6. I considered myself quite lucky to get a comparatively high 3.1. Got a 3.8 in a gunnery test which was much the same type. Questions like what is the best solution with which to clean a gun, what color are tracer bullets painted, and what is the standard service rifle. Very unessential, if you ask me. But so are a lot of things around here. Celestial navigation is getting a bit complicated. Theory of celestial coordinates—which isn't as simple as it might be. Picked up two new instructors this week. One of them is a prince of a fellow and is really going to be swell.

Formerly studied and taught at Stanford. Doesn't care much more for some phases of the Navy than we do. When you come down to it, there are very few men (officers) who are all out for the Navy. Most will get back into civvies as fast as they can. As the Navigation instructor said today, "A Navy instructor spends half his time shuffling and sorting papers. But he has the consolation of knowing that helps win the war." We get some good laughs. One character had to leave class to go to the head (toilet) one day. About three minutes later, he came back, very red of face, and said, "Sir, I can't find it." Takes all kinds of people to make up the world. . . .


December 12, 1943

Dear Folks,
Another Sunday morning, and this one is really cold. A good old Kansas wind, about 25K, is blowing the dust and leaves, and the thermometer is probably down around 30°. Sang in the choir along with the rest of the boys with colds. Everybody has a cold, and we're hoping we'll get to wear our green windbreakers now. They're considerably warmer than the little blue jackets.

Our proofs came in Friday, and I was quite pleased at the results. For once I got a decent "serious" pose and so am having them made, in time for Christmas, I hope. . . .

Incidentally, in the next package from home, I want my white scarf. Anything will help to keep me warm here. The temperature at present is 20° F, which is cold. No snow or ice yet, but any day now.

The week passed well enough. Football Friday was rough, and so was the recognition test. Had a most enjoyable lecture in ENS on British Naval History. The instructor taught English Hist. at Stanford, and it was really like a breath of fresh air. History may be dead, past, and buried, but it has its points for me. I don't see quite how a person can do much thinking, that is, constructive, about present problems unless he has a pretty good idea of what has been tried and thought—how things got the way they are. Then if you have any good ideas you can try them, avoiding past errors. You can't do much for men when you don't know how their minds work. Anyway, that's something for future consideration.

Your letter came today—Dad, I'll do my level best to duck next

Cadet Rea in blues, Iowa Pre-flight, 1943

time. Also today came a very big box of sugared nuts from Berea, Ky., from the Schofields and $5 for a call home Christmas. Tried again to call you this evening, but no luck. Better next time. Next week the Station Chamber Orchestra is playing a concert Sunday afternoon and I'm planning on a pleasant hour. They're really excellent musicians. . . .

This hasn't been such a hot letter. I've been pouring them out lately and am slightly run down. Checked at the U.S.O. Saturday night on connections with Wichita. I'll make it home on my leave in February, but you had better have dinner cooked, and the door open, it won't be a very long stay. . . . The current scuttlebutt is 1. We will get a two week leave. OR 2. We will be here four months. That's what I like about the Navy, It's so definite!

December 14, 1943

Dear Folks,

Not much to write home about. Everything goes on in its own set pattern and is at times bordering on the ludicrous. I fail to see why the Navy spends so much time training men when the job could be accomplished so much quicker. But then I am a little under the weather and inclined to find fault with things this evening.

Speaking of weather, we're having a bit. Last night the first flurry of snow, about a half inch. Today has been cold and the weather prediction for tonight is 20° *below* zero. The worst part is that it doesn't warm up as the sun comes up; it just furnishes light and no heat. Consequently, the cold is about the most penetrating I ever felt. It goes right into the bone despite sweaters and coats and gloves. Ears and noses just freeze. As far as I can see, the Navy doesn't care about the preservation of health here. We are still mustering in our light blue jackets.

There are consolations. We've been dodging ice-cubes in the swimming pool since the first day here, and recently, with colder weather outside, it has been rather uncomfortable at times. The officers in charge make no effort to comfort us (which earns them many epithets) and so we had almost grown used to it. Today, however, the class was visited by the Captain. Now around here, the C.O. is really top dog and carries both prestige, seniority (he was a destroyer man in '17–'18), and rank with full authority. The "old man" had on a big leather fur-lined coat, but after a couple of minutes he started waving his arms and yelling, "It's too cold in here. Why isn't the heat on?" Well, the instructors started running and explaining like mad. The blowers were turned on and continued the rest of the period. The sour faces of the chastised officers were quite a contrast to our broad grins as the temperature began to rise. The anti-climax came later when the Captain was safely gone. Our pet peeve instructor [had] apologized for letting us get cold, and so we swam the crawl and breast stroke at top speed for the next twenty minutes, with promises of more tomorrow. Verily, it takes all kinds of people to make up a world, and it's a good thing the spread is fairly even. We have always been told, of course, that you can't give orders until you've learned to take them. The more I see of the Navy the more I agree. From taking orders from a man you learn quite a lot about him that isn't always apparent on the surface. I have always found that the best officers are the best-liked, and that efficiency is in no way impaired by popularity among the men.

Our sports instruction this week is track and soccer, which is actually enjoyable after last week's "blood and guts" campaign. Soccer was supposed to go outside, but the current freeze has kept it indoor, which is also nice. . . .

There's quite a mystery floating around in the form of free Hershey bars for all hands. Lots of scuttlebutt about the source of much needed (and enjoyed) energy food, but no one's appetite is curtailed for lack of definite knowledge. . . .

December 16, 1943

Dear Folks,

Another day, another pay-day. Fifteen dollars this time—big money. The fourth week is about over, one third of our stay, and I am very tired of the whole place. We are beginning to see why the fellows who come out of Preflight look a little older and tireder than those who come in. It's a grind such as I've never seen before. None of it is too hard, but put together, it saps your energy and dulls your spirit, and with this beastly cold every time you step outside you sure wonder why you're here.

Today I swam the 50 foot underwater swim without too much output, so I guess I don't have to worry any more about that. All studies are coming along well. Have already sufficient proficiency in communications to leave.

Had a little dental work done yesterday morning. A combination of popcorn kernel, hot coffee, and cold air cracked off the back end of a lower left molar. It didn't hurt any, but I figured I had better have it looked at. The head dentist took care of me, filled a cavity in the tooth and built it up as good as ever in record time. Every man gets a dental job before he leaves—usually just before. I watched one poor fellow sit through half of his job—4 fillings and an extraction. My teeth are in pretty good shape and I'm slated for just one filling. Suits me. . . . That's about all for tonight. I have some navigation to do, having already spent half an hour explaining it to my roommates and buddies, and also some high-powered button sewing to do. So write soon.

December 21, 1943

Dear Folks,

Just a note before we go out star-gazing. The pictures finally came and should arrive right behind this—and probably after Christmas, too. I sincerely hope you like the results. I do. This thing of having an hour and three-quarters of swimming each day is amounting to a little work. At least I have no doubts about being able to swim fairly well when I leave here. Am getting back in stride in Navigation. Had a 4.0 average for the last two weeks, which brings my total average up to 3.8. If only my one fouled test hadn't counted so much. Our new schedule is pretty good. Reveille at 0600 (worst part), chow at 0620, academics 0805–1015, PT 1155–1500, Military 1520–1745, chow 1815. Lots of time to relax—before P.T. but none between noon chow and P.T. . . . Caught a glimpse of the turkeys lined up for Christmas chow, tonight. Guess I'll be here for dinner. . . .

December 23, 1943

Dear Folks,

Well, the Christmas holidays are about on us. Saturday schedule (no P.T.) tomorrow, and all day Saturday, and then Sunday. That's going to be one of the best holidays I ever spent. The week has been delightfully quiet and easy in all respects. Swimming and wrestling have been interesting and easy. Today I got off sub-swimming. May regret leaving it, but have gotten a lot of good out of it. But I figure that eight consecutive weeks in the pool will be enough. . . .

December 26, 1943

Dear Folks,

Christmas has come and gone, and, everything considered, I guess it was a pretty fair one. Not having a date for the Regimental ball Christmas eve, I filled that liberty period with fried oysters and a couple of shows. Christmas day a group of us went downtown for waffle breakfast. Only one restaurant was open and we had quite a time getting served. I had wanted very much to call you, so I got a call placed at 0800. I was quite pleased when I got a wire to Wichita in 30 minutes, but no one answered the phone. So I cancelled the call in favor of breakfast. While I was gone, I understand, the call went through again. At 1045 I again placed the call and stood by until about 1400. In the meantime I took time out for dinner—and what a feed. Honestly, I couldn't have gotten better any place but home, and I don't believe that this year I could have gotten as much there. I had practically a whole turkey breast, mashed potatoes and gravy, dressing, cranberry sauce, rolls and butter, green peas, fresh lettuce, and for dessert—hot date pudding with lemon sauce. It was everything you could ask for for Christmas dinner. Incidentally, today's ham and yams and mince pie was no slouch. Well, at 1400 my call reached Wichita again—and again no answer. So, leaving a call for 1700, I wandered over to the union building and listened to their record collection for a couple of hours, including the Beethoven string quartet that Bill, Bob, Lee and I used to play so often and so badly. After hearing it well-played, I can see why we weren't too hot on it. Coming back to the telephone exchange, I was chatting with the operators and learned that any call starting for Wichita at that time wouldn't be apt to reach there for 14 or 15 hours. I threw my hat into the air and sent you a telegram which I trust you received. Will keep trying, though, and may get a call through before I get leave to go home. . . .

January 1, 1944

Dear Folks,

Here I am starting the year out right by standing a watch. Can't say that I mind it particularly, except that it's Saturday and I miss out on three hours of liberty. . . .

Very little news out of Preflight this week. All subjects continue to be comparatively easy and dull. We have a new navigation teacher, but he is no improvement over the first one who was a whiz. Have completed the "study" of naval history (my grade 3.9). Most interesting part was the period spent on the British victory at Jutland, in 1916, when the Grand Fleet lost twice as many ships and men as the Kaiser's fleet and yet won a victory. We've finished off the American navy now and are ready to start on the British in Recognition. P.T. this week was swimming and boxing. Passed my 50′ underwater test, tired swimmer's carry, and stroke test without any trouble. The boxing was most interesting. The instructor spent about two hours explaining the theory of blocks, left jabs, etc., and then we went into the ring for a round. However, the idea in the ring is not to see how well you absorbed instruction, but how well you will stand up and slug and be slugged. I personally don't see much point in being hit more often than necessary, so was cautious of my opponent's fists, seeing to it that hands and arms took most of it. This signalled me out for the signal honor of a second round with explicit instructions to keep my head up and in range where it could be hit. Ah, well, anything for good old Preflight! We put on a fine show, satisfying the deities that we were made of that strong stuff instilled in Naval Aviators. Having a morbid sense of humor at times, I rather enjoyed it, didn't get hurt, I didn't say hit, and learned a little bit. So everyone was happy and P.T. for the week was over. I think I might learn to enjoy boxing a bit if we had a chance to practice a bit of the theory instead of just going out for a slug-fest. But that's the Navy way.

Which brings to mind that I am sending you a little book written by a couple of fellows who went through the program about the same time Bill Roy did. The description of a cadet's career is priceless in spots and altogether accurate. I think you may get some enjoyment from it, as I had a number of belly-laughs from it.* I'm also getting a great deal of pleasure from a Pocket Book of American short stories which Bill L——sent in his Christmas package. It's the first reading I've done, outside of occasional *Colliers,* in a long time, and the change is most enjoyable. Short stories are the only practical thing to try to read around here because you're lucky to get over ten pages read at a stretch.

Everyone is trying to figure out a way to get the 18 day leave. According to the best scuttlebutt, the high men will go straight to Primary with 4 days leave, and the lowest men will be held over for two weeks with leave. The question is, can one safely scuttle his grades until he reaches the desirable position at which he'll get the leave? The California boys are hard up for leave and are counting heavily on it. The safest guess is four days, but I wouldn't be sorry if the two weeks were added. Anyway, we'll see in about four weeks.

---

* Charles Spalding and Otis Carney, *Love at First Flight* (Boston: Houghton Mifflin, 1943).

January 2, 1944

Dear Phyl,

Here I sit. Mainly because there's a chair between me and the floor. I have been writing to people I ought to write to for the last hour and a half and trying hard not to stare dreamily at the Varga calendar on the desk (gift of Bill L——) in between letters. So why not dash off a line to someone I want to write to and would really like to hear from? O.K. We'll do that very thing. Here goes!

Dear Phyl,

So this is 1944? Doesn't seem much different than any other week-end, except for Saturday, which I spent on my feet, standing watch for eight hours. Nothing like starting the year out right—at least that's what the Navy thinks. Good old Navy! The signalman who sent the fellows semaphore Saturday morning took great delight in painting with his flags the joys of a gay and riotous New Year's eve—as enjoyed by every man in the Navy—except the av. cads. Good old Navy. But how can I be harsh when—the gods willing—the same old Navy is going to grant me a two week leave at the end of Preflight. That is the latest and straightest scuttlebutt—and by all that's holy, it better work out. If it's four days, you'll hear from one cadet that is really—well, let's skip the Navy term in view of decency, and say, provoked.

Life goes on here at Preflight with glorious repetition. We've now had one week of everything in P.T. (two of swimming) and are starting return engagements. I'm sleeping through most everything with passing grades and have about as much enthusiasm and ambition as a hibernating gopher. Or do they? . . .

Biggest news item of the recent weeks is the issuance of regulation G.I. dog tags to the Preflight cadets. We now tinkle like any one of Uncle's stalwarts and have one more item to remember when dressing for Sunday inspection. If you're interested, my blood type is "O," and I prefer Protestant burial services. Thanks a lot.

January 3, 1944

Dear Folks,

The first day of the week is over, and for some odd reason I am in the very best of spirits. We have a new schedule again—6:30 reveille—

military followed by academics in the morning. Late chow and an hour of rest afterward. P.T. at 1500 until 1800, and evening chow at 1900: very nice. This week's athletics are swimming and wrestling again. Our test in swimming this week will be a half-hour endurance test and 75 foot rescue carry. Neither is too tough. For competitive sports I am staying on gym, much to my delight. I enjoy the stuff, even if I'm not too good, and I can get a good workout and enjoy it. Which is rare around here. . . .

Leave is the main topic of conversation at present. The fellows in the 38th Batt learned about theirs today and have been telling us all about it. (They are the ones who left Lawrence early.) Everyone is getting a week, and most of them two weeks, but they have a choice for Primary of only Ottumwa [Iowa] or Minneapolis. We are confidently expecting even better. Learned today that train connections from here to Wichita are such that you can leave here at 2000 and reach Wichita about 1200 the next day. It all may not and probably will not work out as we hope, but thinking about it and planning it is the most fun we've had in a long time. . . .

January 6, 1944

Dear Folks,

A short post-breakfast line. It's now 1300. We've had about one inch of snow and ice, and the Calif. and Louisiana boys are going to more trouble to fall down than I ever went to standing up. They're just like a bunch of six-year-olds. Having a lot of fun in swimming at present. Used to think it would be difficult to swim with clothes on, but the Navy has it figured out so that you're crazy to take them off. Just button collar and sleeves and blow a big air bubble, and you'll never sink. It's amazing, and it works beautifully. Very amusing to see some hundred cadets floating around in the pool with humps like hippopotami. Tomorrow we take the half-hour endurance test, which should not be exceedingly difficult.

Glad to hear that a box of hermits is in transit. So much food was on hand Christmas that you couldn't give it away. A week later we had to throw away a lot of it. Now, of course, we're out and hungry.

January 9, 1944

Dear Phyl,

I am sitting at my desk with a big foolish grin on my face, and merely because of a letter from you. Wonderful what it will do for a guy to hear from the right gal. I came in from an afternoon at the movie (Bette Davis in "Old Acquaintance") hoping against hope that there'd be a letter from you—all set to feel sorry for myself—and lo and behold, here is a big fat envelope with the familiar handwriting, and everything is O.K. . . .

Speaking of resolutions, etc. Our room has also turned over a leaf in view of the possibility of going home for a while in the near future. Friday night we installed a swearing box, 5¢ a shot, the kitty to go for miscellaneous pogey bait and geedunks (try that on your dictionary) Sunday evening. Results were quite satisfactory—both monetary and in elimination of profanity.

We've been having a lot of fun swimming this past week. Various endurance tests including swimming with clothes on—which is quite a stunt. All else goes on in the usual way—except that I'm growing careless in regard to my navigation—a thing I must take steps to correct in the immediate future. Three more weeks and we'll be on our way out. Yippee!!!

January 10, 1944

Dear Folks,

Another week has rolled around, and I don't see how they manage to go so slowly. The weekend was enjoyable enough, a show and haircut Saturday night, church Sunday, a concert by the Preflight Concert Orchestra—enlisted men. Very enjoyable. Put on in the Union lounge with no pretense of formality, and a very sizeable crowd.

A very nice weekend, but boy, what a blue Monday this is. I guess everyone has them once in a while, and I had too good a week last week to carry over into another week. . . .

A letter from Bob P——. Just finished ten hours in a Cub. Had a very poor check ride and is quite disappointed about flying. Thinks he'd make a better navigator than pilot and would just as soon.

Finished First Aid Saturday with a big final and today began on what is called Seamanship—a sort of salting course. Well into the British fleet in Recog[nition]. . . .

January 11, 1944

Dear Folks,

Just finished an half-hour of brilliant conversation in Spanish and French with one of my roommates who is about as fluent as I am. Very amusing and surprising how much you can get said—especially with the aid of gestures.

I was always noted in college for being in everything—good and bad, I seem to be keeping it up in the Navy. First it was sub-swimming, and now it is a select group known as Tree Squad. Details of why and how you get on it are unknown and information lacking. The general idea seems to be that it's for cadets not overly-inclined toward athletics. We are supposed to get a good workout of an hour every day, instead of competitive sports. Suits me, and so far has been just that. It's a funny group—one fellow can do 15 chins, which is plenty good, several can do the required 20 push-ups with good form, one fellow is built like a horse and about as strong. No one knows why we're there, but it's "for our own good" and the Navy doesn't need a reason to do anything.

This is the last week for swimming and boxing and everyone will be glad to see them pass. Test in swimming Friday is half an hour with clothes on. That's one I'll pass. It's quite easy once you get your shirt and pants filled with air.

Monday the 38th Batt starts on leave so some of us may be on the way out in two weeks. More rumors that Los Alamitos, Calif., base may re-open for us. . . .

January 13, 1944

Dear Folks,

To quote R. R. Rea of Frankfort, Ky., the slogan of the Reas is "Never a Dull Moment." A good idea and one I'm following through to the letter. I always like to see all of a place, so I'm now seeing the interior of the University Hospital, with the label, "cat" fever. Had chills and fever yesterday, and P.T. was entirely too rough, so I went home and pounded my pillow for 12 hours. Awoke this morning so weak I could hardly get up—so here I am. Cat fever is a very common ailment and not at all serious—a combination of head cold, sore throat, and alternate chills and fever. The cure seems to be rest—and I can go for it in a big way. At present, I'm on a liquid diet, which doesn't go too well. Will probably be up in a couple of days or four, as good as ever. The Navy Ward is very satisfactory—about 46 beds and half of them filled—mostly the same thing I have. Nothing serious—so don't worry. I feel comparatively all right and am hospitalized for my own good. The windows look out on a landscape that might be Kansas, with just a corner of the stadium showing. One of the Wichita boys is just two beds away and I know several of the others. We have a trio of crazy corpsmen around most of the time, and they do a whale of a lot in sustaining morale. . . . Hospital beds were not made for writing desks, so I'll stop this illegible rambling and write again tomorrow.

January 14, 1944

Dear Folks,

It's now 1830, and having some tomatoes and baked potatoes, roll and butter, milk and half a pint of ice cream under my belt, having chased the swing fans from the radio by my bed and found a nice piano quartet, I'm in excellent spirits for a while.

Today was much like yesterday, with a few more pills. I've taken more pills in the last two days than in the preceding twenty-one years. I still don't like them. So far I've had three kinds, one Sodium bicarbonate, one sulfathiazole, the other is still a ringer. This, sleep and rest, gargles and Kleenex seem to fill the bill. As for results, I can't say much. Temperature still up and occasional chills. We have three more cases of cat fever since yesterday afternoon, so you see I'm not alone. The cases range from a couple of men who just came here, to Preflight, that is, Tuesday, to an old chief storekeeper with about six hash-marks.

Busy day today, four medical officers have been through, with another due tonight, the chaplain dropped up (we're on the fifth floor), a navigation officer who gave finals to a couple of 38 Batt boys, and an

Army Air Corps lieutenant and three female friends, apparently sight-seeing. Among other interesting notes, I learned that our head nurse, a Lt. J.G. (as high as you can go in Navy nursing) was at Pearl Harbor, the Coral Sea battle, and another early encounter. She knows her stuff.

There are now three of us here from [Platoon] F-3, and we were a worried bunch last night. One of [the] fellows from the 41st Batt came down with chicken pox, and the Corpsmen immediately informed us we'd be here for two weeks' quarantine. They really had us going for about half an hour, until the nurse assured us that there would be no quarantine and the fellow was removed from our ward. Most worried were the 38th boys who saw their leave vanishing with them in Sick Bay.

I'm hoping to be back on the job Monday morning and get these last two weeks out of the way. I'll be more than glad to see the end of Iowa Preflight. One nice feature of the Univ. Hospital is its traveling library which came in yesterday evening and left with me a fine assortment of reading material. So I'm getting along fine and could really go for a hamburger.

January 15, 1944

Dear Folks,

This kid is getting well. I really enjoyed lunch and dinner and ate heartily, and since this time yesterday, my temperature has dropped four degrees. I'll doubtless be in another day—I hope so for safety's sake, but feel pretty sure of being out Monday. Had a good night's rest last night, despite being awakened at 0200 to take the usual round of pills. If I remember rightly, I have had five different kinds of pills today. One for a cold, one sulfa for fever, one soda for countering the sulfa while it's in the stomach, another brand of sulfa, and a couple of cure-alls. At least I'm getting a lot of practice swallowing pills.

We had inspection today and it was like all others, marked by excessive waiting. The inspecting officer was a four-stripe flight surgeon who took a great deal of interest in all details and patients, much to the concern of the nurses and corpsmen and the amusement of the in-mates—that's us.

I was most glad to get a letter from home this afternoon—it takes a couple of days to get it transferred from P.O. to Batt office, back to P.O. and over to the hospital.

Half a dozen men went out this morning, and most of us are a lot better. The old chief, whom I mentioned, looks like a beardless Santa Claus and has been cavorting around quoting salty rhymes like a fresh boot. (Time out while I take a pill.) So all goes pretty well.

January 16, 1944

Dear Folks,

A lazy Sunday morning in bed, just what the doctor ordered. From the inside, it looks like a good day outside to be inside—cold with an ugly touch of haze that would be pretty in Kansas. I hope like everything I get out of Iowa for Primary. The people are all right and the land good enough, although any of it I've seen yet would be pretty poor in some parts of Kansas. I guess it's just Iowa—that I feel about it as the Calif. boys felt about Missouri—not quite up to par. . . .

I am apparently about the same and plenty willing to get up. One of our corpsmen came down last night with a chill and temp[erature] of 105°. The worm turns! The baby ward seems to be somewhere near, and why, in the name of mankind, didn't the fellow who invented these bendable beds bend them at the head instead of at the foot, and arrange some means of lowering them to a 0° angle instead of 5°. I'm very fed up with up-hill sleeping. No sulfa today and chicken smells wonderful, so I must be getting better.

January 19, 1944

Dear Folks,

Am now out and O.K. Feeling kind of weak and woozy, but much better just being up and around. Am very busy and rushed keeping up and catching up, so won't have time to write more even if there were more to write. We're having our navigation finals next Thursday, and they're going to be pretty stiff. Will really have to cram. Some of the 38th Batt are going to Hutchinson [Kansas] for Primary, oh well. Write soon—

January 22, 1944

Dear Folks,

Well, here's the straight dope. As I wrote, I left the hospital Tuesday morning and Wednesday was feeling pretty fair, attending all classes and getting a light work-out in P.T. plus a heat lamp. However, Wednesday evening I developed a most unpleasant earache that progressed with the night. Three aspirin left me a bit woozy, but afforded a couple of hours sleep. The next day I turned in to Sick Bay and they put me to bed after treating the ear. So Thursday and Friday I was on my back again and not very happy with the world. The original ache has disappeared, but it still is swollen and about as useful and comfortable as a head of cabbage. I get a treatment each morning which, I hope, will clear it up soon. In the meantime I have to keep it packed with cotton, which isn't pleasant either. Dad's letter came Thursday morning and was really a help as I was pretty close to down and out. However, this evening I literally girt my loins in Navy blue and made it to town for a fruit plate supper and am now back preparing to study like blazes. I'm so far behind it is entirely possible I may be dropped back a Batt, but I hope and pray and work to the end that that shall not be. Two extra weeks, when our Batt goes on leave a week from Monday, would be a punch below the belt. At that I'm better off than one fellow who is having to have three front teeth pulled. . . .

P.T. this next week is labor engineering, so it won't be rough. If I can stay out of trouble now, it'll soon be over. Local news—Capt. Hanrahan died yesterday afternoon. A real loss to the men and cadets— a true friend and officer. Don't worry about me—I'll pull through with flying colors. After a week out, I'm still the fastest navigator in the bunch—and boy, I'm navigating homeward, and that's a true course.

January 22, 1944

Dear Phyl,

I am late in writing, but as I've been flat on my back since you called, I've been unable to put pen to paper earlier. As I told you, I spent five days in the hospital, and then got out two days, and went back for two more. I am hoping and praying that will be the end of it and that I won't be set back a Batt because of time missed. Some of our men go on leave a week from Monday, and it would be rough to have two more weeks stuck on me.

Your call was as unexpected as it was enlivening. It was like having some distant window opened and a shaft of light break through into the outer darkness in which I stand. The sound of your voice moved back the walls of reality to a nameless time and place where people are humans and not machines, in which men and women live, love, and work without the overcast of fog and clouds and the turbulence all around which is our life today. In the normal run of things, it will take a long time for that to return—and who am I to talk who have but fought the battle of Iowa Preflight? It is strengthening to have this dream revealed, if only for a moment, and for that one who opens it unto us, we name a special star and wish upon it each night.

January 25, 1944

Dear Folks,

Tuesday and I'm still carrying on as usual. Not enjoying life, but each day passed is one nearer the end, and I intend to see it through. Finals begin tomorrow and end Monday. I'm feeling pretty good most of the time, but toward evening I get to feeling kind of lousy.

As I mentioned, Capt. Hanrahan died Friday, and the services Sunday afternoon were very impressive. No successor has been appointed as yet.

Am awfully tired and have a big recognition final tomorrow, so will sign off and hit the sack. . . .

January 27, 1944

Dear Phyl,

Talk about rushing! No time to think, much less to write. Am feeling O.K. now. Your call helped tremendously. Having finals now. Cracked the navigation final today. Four more tomorrow. All signs point to my leaving Monday—am definitely "on the list." Won't believe it till I see it. Will try to write a *letter* over the weekend. Taps in 3 minutes.

January 28, 1944

Dear Phyl,

At last an evening and a moment to write. Today we had four finals. Yesterday the first big navigation quiz and two make-ups. One on Wednesday. One more in nav. tomorrow and in Aerology Monday. And if I'm lucky, I'm through Preflight. The first Nav. quiz I cracked 3.9. Am reasonably sure of passing tomorrow and should do well. An E.N.S. test today has me worried. If I flunked it, which is within the realm of possibility, indeed too probable, I'll have sunk a sweet set up. To wit: I am "Sat," that is satisfactory in everything, am due to leave in the first draft Monday on nine days leave, and report for primary at Olathe, Kansas. That's about all I could ask for. I'm praying for some good guesses on that test, and you can see why. It was a tough test and I missed out on much of the material covered, so I'm not predicting an outcome one way or the other. That's the news from U.S.N.R. . . .

I seem to be getting back to normal—outside of 11 pounds lost and a bad ear which still makes me feel like a post, and am hoping that will clear up soon. The *"if"* week at home will take care of the former item, I know. It has been a peculiarly interesting experience. I haven't been on my back for any length of time since heck was a pup, and the results are surprising. I think and work just as well as ever, but when it comes to any creative thought—such as letter writing—I just haven't got it. In the last two weeks I've written home once, no, twice, and you twice, and that's all. It's something of a record. Even the idea of being home next week, in less than 96 hours, doesn't give me much of a lift. Just getting along from hour to hour seems quite enough to satisfy any and all "inner drives." All of which seems to me the picture of one very devoted and tied to himself, who needs jarring loose with a good kick in the pants. That's about the effect your call had, only unfortunately, telephone conversations must end—usually after five minutes. Well, that was wonderful, anyway. The week has been one that will leave me in fear of no college finals week, past or future. Maybe I'm just tired.

# 5

## Spring in the Air

THE KANSAS WEATHER that Bob Rea described during his Primary training followed a familiar pattern—from snow on his arrival at Olathe to tornado warnings and finally just plain summer heat. But if expected, it was not always predictable. The war was shaping up the same way. In England, as the censor allowed *Life* magazine to reveal in an array of photographs, the stockpiles of war matériel were vast and breathtaking in preparation for OVERLORD.[1] The general public—and the enemy—knew an invasion was coming but had no real idea when. OVERLORD was made possible by a series of events in February and March 1944, when the tide of the air war turned as the once proud and aggressive Luftwaffe suffered a massive defeat. In February, Allied bombers concentrated on enemy aircraft factories, and in March, AAF heavy bombers joined the RAF in attacking Berlin. In the ensuing daylight air battles, swarms of far-ranging AAF fighters shot down the cream of the Luftwaffe's air defense force. On D-Day, June 6, there were virtually no German planes to be seen during the crucial hours when the invaders came ashore, and this situation prevailed during the next forty-eight hours as the troops moved inland from the beaches.

There were also signs of an approaching storm in the Marshall Islands in the Pacific. They appeared in several forms: a fighter the Japanese called "Whistling Death," the inverted-gullwinged Corsair; the stubby Hellcat, which would destroy more enemy aircraft than any other

[ 132 ]

U.S. Navy plane; the sleek torpedo-carrying Avenger; and the Helldiver, which had begun to replace the old Dauntless, the veteran of the Coral Sea and Midway. Their bullets, torpedoes, and bombs fell like hail and lightning on enemy transports and naval vessels. No Japanese planes remained in the atolls to welcome one of the mightiest armadas yet assembled. The guns of the U.S. fleet thundered without letup as LCIs (Landing Craft Infantry) emptied assault troops on the beaches. Kwajalein fell, then Eniwetok. Carrier planes went hunting the enemy fleet in its lair at Truk in the Carolines. It had gone, but the attackers' frustration cost the Japanese almost 300 planes and many auxiliary vessels. As in the case of the Germans, these losses were increasingly hard to replace. All in all, February 1944 was, for the U.S. Navy, a very productive month. The next targeted islands in the Central Pacific, the Marianas, would provide ideal bases for a new AAF superbomber being readied for operations against Japan.[2]

If the action in the skies was exclusively male, that did not mean that American women were not engaged in the war effort. True, they did not fly planes in combat, as Russian women did, but American "girls" ferried warplanes and aided in gunnery training of male aircrews. In a letter home, Bob Rea acknowledged the professionalism of the female naval personnel, the Waves, at Olathe. Rosie the Riveter was a household name, and *Life* magazine, harking back to a legendary American feud, featured two young women working side by side at a war plant in West Virginia, representatives of millions of women essential to the "Arsenal of Democracy"—one was a Hatfield, the other a McCoy.[3]

For a Naval Aviation Cadet, life began anew at his Primary Flight Training Base. Here he flew the renowned and beloved Yellow Peril— the N2S Stearman biplane, which symbolized the legendary military aircraft of World War I, the great flying movies, and all that had gone before this increasingly high-speed, supercharged, radar-controlled war. With the wind in his face, the wires whining, a white silk scarf streaming out behind, every cadet became an ace. Happily, he did not know the sad fate of most of those gallant knights of an earlier war, though some learned it the hard way.

Olathe existed for flight training, and flight requirements imposed their own discipline. It mattered little how snappy the salute, nor did anyone cry "Hit the bulkhead!" when an officer came down the passageway. On the flight line it was another matter. Hearts beat faster and, on gusty days, wingtip accidents were frequent. Instructors were both tutors and terrors. Their favorite trick was to switch the gas tank control to "off" just as a cadet began taxiing out for takeoff. The wise cadet watched that little red handle more closely than he watched the traffic pattern. Lieutenant (J.G.) McDuffie, the acrobatics instructor, was a cheerful madcap who demonstrated that a Stearman could take more than a student pilot could enjoy, and he only regretted that the faithful Pratt & Whitney radial engine was not equipped for inverted flight. Mac was a good friend and teacher—even if he very nearly spun in one morning when competing with his student to see who could make the most precise S-turn small-field landings.

It was something of a culture shock to encounter free keg beer at a station party, but the war put an end to prohibition in Kansas. On the other hand, the war encouraged romance; friends became engaged and married, but cadets were forbidden to marry before commissioning— and judging by the rising attrition rate, the Navy seemed to have enough pilots. Such concerns disappeared with leave, however, a time when proud cadets discovered that starched white uniforms went ill with the tan liquid stockings young ladies applied in lieu of nylons—one of the lesser hazards of the service.

## Primary Training, U.S. Naval Air Station, Olathe, Kansas, February 9–June 24, 1944

Wichita, Kansas
February 2, 1944

Dear Phyl,
    . . . As you note, I am actually at home and out of Preflight. It is a wonderful feeling to be able to look forward to the next day and the next base—Olathe. I'll be flying again, and that's good. Am spending my 9

day leave seeking ration points, seeing a few old friends, and resting. Few are left. Most should [be] or are leaving.

February 9, 1944

Dear Folks,

Arrived at 0730 and got a ride up to the base without any trouble. Logged in, got my bunk assignment, linen, locker, and unpacked. Ran into a former roommate at Lawrence, and we gave the base the once-over. This afternoon we've filled out the usual routine forms. It's quite different from Preflight—you meet but two musters a day, go to chow on your own, may go to a picture show every night. . . .

The Olathe Air Station is built on good old flat Kansas prairie. Only visible trees are the barren hedgerows. Buildings vary from brick to board. We are living in real barracks. The entire cadet regiment in one building. In our barracks sleeping rooms are about thirty double-decker steel bunks—the best we've had so far, but nobody is very enthusiastic about living out of a locker with sixty other fellows. However, there seem to be plenty of advantages to overcome this. There is a fine recreation building with full facilities, stores, shops, theater, etc.

Planes are thicker than flies at Liberty; it's nothing to see fifteen N-2-S's coming in or taking off at one time, plus an SNJ or Skytrain tearing in at an odd moment. Got a glance at the syllabus and it's plenty tough—not at first, but they cover everything. Stage A is mainly soloing. Stage B calls for precision and accuracy, Stage C is aerobatics, Stage D night flying and formation. The whole course is supposed to last eleven weeks, but is lasting longer due to bad weather. The fellows who have been here some time say the ground school is quite easy: navigation, aerology, communications, gunnery.

A good part of the work around the base is done by the Waves. They serve as mechanics, yeomen, personnel officers. Ensigns and L[ieutenan]ts. are thick but not held in the awe which elevated them in Preflight. We will begin flying Monday.

Liberty seems to be plentiful in comparison to Preflight, but do not count on my getting down to Wichita. Although it is a bit early to start counting on it, there seems a good chance for some leave at the end of the course.

Big news is that Hutchinson [Kansas] is closed as a Primary base. All the fellows assigned there were sent to other bases, so I guess I'll not be able to offer any hospitality to my friends. Have no idea where they were sent. Some of the Hutch. draft came here—others went to Minneapolis.

Noon chow was quite good, in contrast to a late breakfast. Reveille is at 0630, so I should get plenty of sack time. This about covers the situation at present, and I'll write more as it happens.


February 10, 1944

Dear Phyl,

We are apparently free for fifty minutes, so I'll see what I can do on a letter to you. The leave ended much as it started. The week did rest me up a lot, though, and I feel ready to give this base a good run for its money. It's a flat, treeless collection of barracks, hangars, and miscellaneous buildings, at present covered with a thin, mushy layer of snow. Our quarters consist of one large room bunking about sixty men with lockers for gear storage. Quite a change from Iowa. The one nice thing is that the bunks are new and good. We are getting Navy chow now, and while far from fancy, it is very good. Planes we have in abundance. It's nothing to see half a dozen planes coming to land at one time, and the air is always full. I got a glance at the flight syllabus, and it is plenty rugged. They cover everything and in detail. Ground school, it is rumored, is pretty much of a snap. We shall see. Military discipline is back on a sensible basis and the sight of an ensign's stripe doesn't call for a dress parade.

We are at present undergoing the first half-week of indoctrination, which is a necessary evil and always a little hard on morale. Next week we begin flying and working in earnest, and life will pick up. The base has a fine recreation building—bowling, pool, ping-pong, barber shops, ship's store, Waves' lounge (for dancing), theater, with a show every night and a new one every other day. Kansas City is close and available. . . .

Later: This is a most amazing place. We've been here two days, nobody is excited, we've had only three hours of indoctrination, one period yesterday afternoon, and one this morning. The rest of the time has been free—a novelty for the Navy. Oh, oh. I take it all back, the schedule for the rest of the week just came in—we'll be busy enough. It's quite amusing to walk down the passageway with your cap off, meet an officer and ignore him completely. Poor old Preflight—they tried hard.

Scuttlebutt concerning the washout rate here is rather vague. Anywhere from 50% down to 15%. H—— L—— washed last week, from this base, but has a chance at midshipman's school, I hear. No fatal crashes lately, either, which is good news. Yesterday we filled out lengthy "casualty" forms—just in case. . . .

February 11, 1944

Dear Folks,

The indoctrination days are about out of the road. They have been the easiest I've seen. Usually this first half week is pretty much of a headache, but this has been enjoyable. So far we have checked out athletic gear—the usual; flight gear—helmet, goggles, gloves, mittens, a swell jacket—and sundry books for ground school. The air of relaxed discipline holds, and all the people, instructors and what have you, are interesting, intelligent, and treat you as equals. Very nice. Even the barracks don't look so bad, despite a freezing rain and snow flurries. Passed the physical exam today without any trouble. Will have to have a little dental work, however. The ear will get examined tomorrow. It has been giving me no trouble and at times seems to be back to normal. Classes begin Monday. Navigation will continue to become more practical. Aerology also—met the base aerologist, a two-striper [Lieutenant] with two rows of service medals—including the Purple Heart—and time in the Aleutians. Really knows his stuff. Communications is in two parts—radio, visual signal, telegraphy methods and procedure, and the usual code and blinker at 11 and 8 words/minute respectively. All I should have to do is bring my blinker up two w.p.m. to check out with ease.

One of the fellows has a very fine portable radio and phonograph, and it is a pleasant change to hear Tschaikowsky, Beethoven, and Verdi along with the usual stuff. Right now John Charles Thomas is crooning "Di Provenze" from "Traviata," and in another corner Harry James and his trumpet are calling up the gremlins.

We are to draw our travel and subsistence allowance tomorrow before liberty, and possibly some more. Incidentally, this is the first base at which new cadets are not restricted for a week or two. Nice!

Chow is good—and quite different from Preflight. The soup, for instance, tastes like soup. We've had fish, roast beef, spam sandwiches, eggs sunny side up! So I'm eating well. And preserves, to boot. So much for now. Mail call twice a day—how about keeping the P.O. busy. . . .

February 14, 1944

Dear Folks,

Monday morning and snow on the ground. Real blizzard last night—wasn't sure the bus would make it up and around all the K.C. hills. The liberty was very good. Nice to see Kansas City again; it is considerably better than Iowa City. This, incidentally, is the first place we haven't been restricted for a couple of weeks, just on general principles. Getting in to K.C. is another matter. If you're lucky, you catch a ride or get on one of the three buses for a thousand men. Not much fun standing out in the snow for an hour waiting for a bus. The liberty schedule is a bit odd. Alternate weekends an overnight until noon Sunday, and station liberty. Every 10 days another similar deal while your instructor has two days off. I might get home once or twice, but it is very doubtful. . . .

The medical officer examined and tested the ear carefully Saturday morning and said it was coming along fine and clearing up as it should. Hearing not quite normal yet, especially on high vibrations, so he grounded me for two days and took me out of swimming for a week. Go back for an examination tomorrow. It is much better than when I was home and ought to be all O.K. in another week.

I suppose you are busy entertaining another guest this week. I could certainly go for some more of that good food. Although we're eating as well or better than ever, I just can't seem to get filled up, no matter how much I eat. Had two steaks in K.C. and a chop. Very good, very small, and very expensive. Did I tell you that we actually had pancakes for breakfast last week? French toast this morning. Haven't weighed yet, but expect to be back to normal thanks to you and the Navy's chow. I'm really feeling good now, and very anxious to get up and see how the Stearman handles. The flight officers we have talked with so far have a great deal of respect for the old N-3-N, but the N-2-S

N2S Stearman "Yellow Peril" flown in Primary training at NAS Olathe
(Courtesy U.S. Naval Institute, Library and Photographic Services)

is supposed to be considerably better. We have a newspaper here, but like the *Spindrift* at Iowa City, [it] is put out by and for enlisted personnel and not particularly for cadets. . . .

February 16, 1944

Dear Folks,

Pardon the blot. Altitude does that to fountain pens. Got ungrounded today and went up for my first hop. Met my instructor yesterday and the plane today. Both are swell. He is a young fellow, serious, and knows his stuff. The plane is sweet. Much like the N-3-N's but easier to handle and much cleaner. Cockpit arrangement is different, but I should be on to it in a week. The traffic pattern at the main field is quite simple, but most of the 15 outlying fields have individual traffic patterns. There are plenty of planes in the air and landing at one time, but if you keep your eyes open and don't "dope off" there's no reason for trouble.

The N-2-S lands and takes off easier than the N-3-N and is certainly more responsive to the controls. But I'll be able to tell more about the plane and the flying—and the pilot. . . .

Had an entrance exam in Navigation the other day. Boy, you sure lose the touch quickly. But I got through without too much trouble, with 15 minutes to spare and a number of correct answers. Today we get code and blinker. Have to check out at 11 w.p.m. in code and 8 in blinker. The latter will be a little work. Had a lot of fun in gunnery yesterday tearing down a Colt .45 automatic. Total time for stripping and reassembling about two minutes. . . .

February 18, 1944

Dear Phyl,

There are advantages to bad weather—time to write to you. It's one of those beautiful days: not a cloud in the sky and the temperature + 7°F.—and the Navy prefers its pilots without frozen noses and toeses. So we're grounded. Everybody is anxious to get up and solo and start to work in earnest, but I've had enough open cockpit time that I'm not too eager. Had one hop—just flew around the area to get oriented. We have sixteen auxiliary fields and have to know their location and use, as well as traffic patterns. The Stearman is a sweet plane, much nicer to fly than the N-3-N we had at Lawrence. And boy, the way they fly 'em around here. Perhaps you've heard comments about the way Navy pilots fly? Well, these boys are the ones that do it. They make their turns at about 50', climb at a 45° angle, taxi like fiends, and when they all come in over one point it's great sport to see who's going to get there first. This base is supposed to have the highest mortality rate of any Primary station: they have lost about forty instructors and students in the last twelve months. However, most of them were due to pilot error—for instance, one instructor tried to get off the field during a tornado, and six others went out hunting ducks and doing slow rolls at two hundred feet. None came back. But then, instructing is the most dangerous branch of Naval Aviation. That's a fact.

Ground school is coming along in fine shape. Made a 3.85 on the Nav. entrance exam. Aerology will be O.K. and is continuing right where we left off in W.T.S. Gunnery so far has consisted of field stripping a Colt .45. Eventually we get a little target practice. Can already take Code and Blinker fast enough to check out, so if I can fly, this ought to be a good three months. The weather is apt to stretch the intended eleven weeks out to thirteen or maybe fourteen, which would mean we'll be here until early May when the weather should be perfect. . . .

(0800) To show you what a tough life we lead, here's our schedule, which just came over the P.A. system: 0845—go to see a basketball game; 1000–1015 inspection; 1045–1145 picture show. Then chow and ground school. Who's kicking? Not this boy, I can be as lazy as the next guy.

You asked me what I thought about the Waves—and your joining. That's a ticklish subject. There is no question about the relative merit of the Waves. It's the best women's corps in the armed services. The officers whom I've seen and met here on the base are very intelligent and capable young women who are doing their jobs well. The enlisted men who are being replaced by Waves aren't so enthusiastic, and the pride of manhood is severely taxed at the sight of a female who rates a salute after four months' training. Having never read any Wave literature I don't know their stand, but personally, I think a Wave officer has a great responsibility. She is a Naval officer and as such has plenty on her shoulders. She will be in command of men who have spent from four months to fourteen years in the Navy, and it takes more than rank to command successfully. . . .

February 20, 1944

Dear Folks:

Sunday afternoon is always a good time to write letters, especially after a good dinner and a good flight. After sitting around all week, we finally got up again this morning. I have a new instructor who will stick, I hope. We shot landings all period and I was pleasantly surprised with the results. Am beginning to get onto the feel of the plane and the look of the ground from the air. It was a swell day to be up, fairly warm and clear. I see in the aerology office that weather is good around Wichita. Yesterday we had a lot of fog and smoke from K.C., however it looks like we might have some good weather for a while. This week we will fly in the afternoon, which means a much fuller schedule and less time off. Saturday seemed a day devoted to inspections. During the day we had three—barracks, locker, and personnel. This latter was a fairly impressive ceremony—reminiscent of Liberty. The battalion massed in blues in the gym, a close inspection of personnel made (down to ties and caps), and we passed in review to the music of the band which appeared at Liberty several times.

I'm spending this weekend on the base because of limited time and navigating to do and letters to write. . . . We're getting paid again this week and it looks like Preflight days—$10. I'd just as soon get it less often and in greater amounts. . . .


February 22, 1944


Dear Phyl,

. . . Last night I discovered the base library and music room. Both are small but nice and very well equipped—everything from Mozart to Strauss and Shostakovich, Zane Grey to Sam Pepys. Also Mahan on Naval Theory.

This isn't apt to be such a hot letter. I had #3 flight today—half way through A Stage—and it was awful. Sometimes I act as if I'd never seen a cockpit before—always when the instructor is around. A 30 knot wind didn't help either. Both the Ensign and myself were well disgusted. And when I have a bad flight I am not a pleasant guy to have around. So I'll take it out on you. . . .

Tell me, Phyl, how would you feel if I were to wash out of flight training? It is a very possible thing, you know. Flying never meant much to me before, the wings, that is, but here it has taken on a new color and means a lot. One of my good friends at school, a freshman last year, washed out of the AAF recently, and I'd hate like everything to get a letter full of the stuff I wrote him. It's no disgrace, but you've gone through a heck of a lot to find out. Oh, well, I am in the Navy, and I'll take that to anything else.

Enough for this: now to a few of your questions. . . . I hate practically all radio serials as much as I like newspaper comics. I don't play bridge, but am ready to learn any time you think you can teach me. I'm glad you like "New Yorker." I discovered it in the S.U.I. hospital and read about a year's issues. . . .

Dreams or plans for the future? Nothing definite—grad school I suppose. I'd like to make it straight through to a Ph.D. if possible. Probably not. I'd also like to take a year off to read, study, travel, think, plan—that's a dream. My Dad is sixty-some and still trying to find that year.

I particularly liked that Sandburg verse—"To Know Silence Perfectly." It describes my feeling about music—and silence—perfectly. In the depths of great music and in the grandeur of an all-pervading silence, man may look for some reflection of his own puny soul—in it find his small dreams and his great visions magnified and brought to life. Or he may lose himself and the world in which nature and fate have constrained him and know a calm and peace which is to the soul as untroubled sleep to the body. Sometimes I wonder if there isn't some relation between this feeling of a "perfect silence" and death.

This is a most incoherent letter, methinks. Today a new class came in with a number of my mates from Preflight, one roommate and another very good friend. . . .

February 25, 1944

Dear Folks,

. . . This has been a typical Kansas week. We've had wind and rain, fair and sunshine. Yesterday I flew without boots or gloves and was quite hot. It was a perfect day, little wind, just like April or May. Had a very good flight and some encouraging landings. If you handle it nicely, the N-2-S will do about whatever you want it to.

I've been spending some enjoyable moments in the music room. It's really swell to come down from a flight and go over to the Rec Bldg., get a coke and stretch out in a comfortable chair and listen to the strains of Tschaikowsky or Beethoven. Last night's movie was quite a change— a mystery picture entitled "The Uninvited." Kept the barracks awake half an hour after taps trying to get the plot straightened out. . . .

Today is more than typical—we got up at 0630 and went to chow at 0645; there was a touch of moisture in the air. At 0730 we went to PT in a gentle, but wet, rain. At 1215 we prepared to go over to the hangars and foul weather program was announced, to begin at 1245. When we mustered at 1245, flight was announced. So back in, change to flight gear. Now, an hour later, it looks like the weather might close in again in an hour or so.

. . . Got quite a surprise in the show last night. Had a close-up of a

soldier at Coney Island, and if it wasn't one of the fellows who went to
F.U. last year, my eyesight's failing.

We have a new bunch of cadets in—former W.T.S. instructors who
will be given Navy training and commissions. Dressed as civilians, they
are very noticeable. . . .

February 25, 1944

Dear Phyl,

The music room is a delightful place in which to write letters, but
between the listening and the thinking (for I do, even if the letters don't
show it) it's pretty slow work. However, it's by far the best place to
spend an evening. I have just re-read a half dozen of your letters and
feel very guilty about the number I write. I am the envy of the barracks
for the number of letters I receive, thanks to you—and I mean that. . . .

Flight is coming along with the rapidity of a prairie-schooner. No
hop today. Yesterday, however, I had a swell flight. I am beginning, I
hope, to get the feel of the plane again and back to the point where I
can start to work in earnest. We've been having strong winds this week.
Yesterday a couple of fellows cracked up on formation take-off on the
main field. Fortunately they were low enough that no one was injured.
Today one of our fellows—in my platoon—and his instructor eliminated a telephone pole and an N-2-S on one of the outlying fields. The
plane was completely washed out, but both occupants escaped without
a scratch. They were exceptionally lucky to have avoided a fire which
might easily have been fatal.

Wound up swimming yesterday with a 100 yard rescue carry and
a half-mile swim fully clothed. The former was rather exhausting, and
I'd have to like a fellow pretty well before I'd lug him that far or farther
without any outside help. Don't know what we'll have next, basketball,
I hope.

Next week we go back to the morning flight schedule and time will
again hang heavy on our hands. It's sure a swell feeling, though.

My mail has been quite light this week except for your letters. Dad
has been in Chicago at a warehouseman's convention. Wrote that he
was in a $16.50 suite at the Edgewater Beach Hotel and having a great
time. He'd like to retire, but you can't very well just walk off from a very
lucrative business which depends on the man running it, and there is
no one to take his place. What will happen after the war when ten million

young men return to the business world? There are jobs now, but will there be when the government stops building armaments and the "defense industries" return to their normal production.

If ever there was a time of uncertainty throughout the world, in the latter years, we are seeing it—a period of world revolution, the course of which we can't yet see, but only guess at. Russia, China, India, Japan—there is an interesting question-mark. And a Europe from which the name of France has been obliterated—we credit Napoleon for much of modern history; Hitler has not fallen short of the little man's mark. Why expect any stagnation or return to what is our "legitimacy"? We will feel all this in America, and I wonder about the principles we are told underlie this government. If "these are the times to try men's souls," as has been suggested, the trial is not to see if they will stand against the storm, to stand is to perish; these are times to move ahead, to rise, to fly, to cut across the path of circumstance and blaze new trails and find new worlds to conquer. I wonder?

February 29, 1944

Dear Folks,

Another week over and another week beginning with no flight. Snow on Monday, and today and tomorrow are my instructor's days off, so I've been taking it easy this morning. Did a little navigation and reading and listening to a Schumann symphony. It is a real treat to have time to read for pleasure—a thing I've greatly neglected for a long time. I've already gotten thru a Maxwell Anderson book and a biography of Marshal Ney, one of Napoleon's greatest generals, and am slowly easing into Chas. A. Beard's "America in Mid-Passage" which is a 700 page treatise on the years 1924 on.

The weekend was most enjoyable. Went into K.C. as per usual, arriving about 6:15. It was a sort of old-home week for P—— and myself. After dinner I accidentally ran into Ruby W——, who is with TWA in K.C. now, and a friend, also from Wichita. We had a wonderful couple of hours talking about old places and faces and common acquaintances. Came back about 1030 Sunday morning in what was pure soup. Ceiling and visibility were almost zero. So I spent the afternoon in the library listening to records, reading, and writing letters.

Studies are coming along very well. Made a 4.0 on the first navigation test. I learned from P——, who stayed at Iowa a week longer than

I, that I was high man in the Nav. class there with a 3.9 average. Recognition is a little tougher now—British destroyers, and now we're beginning on Jap planes.

Passed all my swimming tests—half-mile swim clothed and 100 yard rescue carry. Stranglehold technique and various forms of artificial respiration. The next two weeks we have tumbling and apparatus work. They have a very eager calisthenics man here. Works us as hard in ten minutes as they did in half an hour at Preflight. Everybody woke up with a stiff neck this morning as a result of yesterday's workout.

We have had quite a run of accidents lately—three in three days, in fact. Thursday, a formation of three planes took off and one wing-man got caught in the leader's slipstream and cracked into the other wing man. No one injured, fortunately. Friday, one of the boys in our platoon and his instructor took out a telephone pole landing. The plane was a total loss, but the student escaped injury and the instructor received only a sprained back. The kid was pretty well shaken up, though, and grounded for a week until his nerves calmed down. Then Saturday, the Lodestar cracked up on the field, which you read about. The K.C. papers were full of garbled accounts—all wrong. The plane had seventeen aboard, with some standing in the aisle, well overloaded. The pilot attempted to climb too fast on take-off, got up about seventy-five feet and she stalled out. He nosed her down, got some flying speed, but pulled up too sharply before he had sufficient airspeed. This time it dropped off on one wing and spun in. Fire broke out at once, and most of the occupants didn't have a chance. The nose broke off and the pilot and co-pilot got out along with two or three others, one of whom later died. I was up at the time, and when we came in the field was closed and we circled over the wreckage. Later I got a close-up look at it. Half of the tail assembly was still recognizable as airplane. The remainder simply disintegrated. Cause of crash was determined quite simply as pilot error.

The cookies arrived yesterday and are being greatly appreciated. The combination is especially good. Thanks for the clipping on T—— and I——'s engagement. Haven't heard from him since I was home, but now that all is settled expect a long explanatory letter. No doubt it was quite a strain on the old fellow's nerves.

Dad would be interested in our latest gadget—a device to locate and name any star of 1st, 2nd, or 3rd magnitude when only the one star is visible. Most ingenious—and practical. Laundry service here is fine, and they actually press shorts. Much easier on clothes than Iowa City, too.

March 2, 1944

Dear Phyl,

.  .  . Life comes and goes here, and at present it seems to be going fast. I had my first check today and got a down on it. I have two more to go and am still confident that I can get ups in both, but it is a kind of a jolt. I'll know the final results in a couple of days. It's going to be quite amusing if I wash out so early in the game. I don't relish the thought. .  .  .

I'm not in a very good letter-writing frame of mind, thanks to this morning's results. I am, so to speak, with one leg in the grave, and I'd like to find out which way the other one is going. I can't help wondering what certain people would think or feel if I shouldn't make the grade. I'm also curious how I'd take it. What do you think, Phyl? .  .  .

March 3, 1944

Dear Folks,

It's a cold, wet day and it remains to be seen whether or not we fly. I am both curious to get up and again not so. Yesterday I flew my sixth period and got a check. Flew what I thought was a fair ride, but my glides were too slow for the check pilot and he handed me a down-check. So today I'm due for another and possibly one tomorrow. If both of these are up-checks, I'm O.K. If not, then I try for extra time and some more checks; if the request is turned down, I'll see something of the sea navy, beginning at Great Lakes. This is perhaps, and I hope, a pessimistic view of the situation, but that is the way the land lies. I am confident enough of my own ability; it's just a matter of proving it at the right time to the right people. It's a lot easier to get an up the first time than the second and third times. Yesterday's check pilot gave two downs in one hour; I'm hoping today's will be in a better mood. At least three fellows in our class have already had two downs, and one is up for board action this morning. This is a rather sad state of affairs, I realize, but all I can do is my best and hope for some luck, which in flying counts for a lot. If things don't work out, that will be tough, but not so much a disgrace as a disappointment. It would certainly be better to wash out now than crack up later on.

So much for that. Time will tell the final outcome. T—— finally wrote, after the strenuous job of becoming engaged, and seems quite satisfied with the job. .  .  .

March 6, 1944

Dear Folks,

Not much news except that everything is O.K. again and I'm off on B stage. Had my second check Sunday morning and despite cold, low ceiling, and bad wind, got a pretty fair ride and, more important, an up. Had a Ground Period today which satisfied me as the wind was pretty strong and very gusty. Fellows were getting wing-tips all over the field and outlying fields. It was a beautiful day, but about as cold as any I saw in Iowa.

Ground school is coming along alright. I have a 3.8 in Navigation, and other subjects are satisfactory. Recognition is pretty rough now that we are studying the Jap fleet and air force. P.T. this week is apparatus work. Mat work last week. Very enjoyable sport. . . .

Apparently there was another crack-up last night—a C-47 Skytrain, some place away from the main field. No details available, though. News from People I Knew:—B—— P—— is in radio school in Illinois, H—— W—— is hoping for a change, T—— wrote a very self-satisfied letter containing details of his recent engagement, M—— H—— is still on the East Coast with AFSC [American Friends Service Committee]— having dinner with Mrs. Roosevelt.

Got paid tonight and will probably forward $20 for safe-keeping soon. You asked about our lay-out here:—a *crude* map.

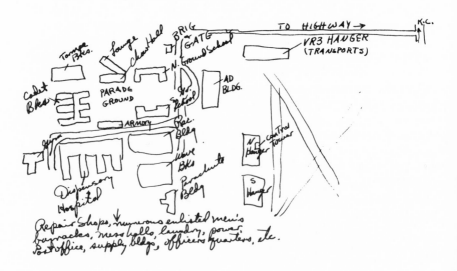

March 6, 1944

Dear Phyl,

I have a vague feeling that maybe you are a bit put out at the scarcity of mail from Olathe, eh? Well, I haven't written until I knew how things were going to turn out. Yesterday they "turned" and I am pretty sure of being here a while longer, having passed the necessary check rides. Needless to say, I am quite relieved. I don't believe I've been so really scared in my life. You don't realize just what a thing means until you are close to losing it, and I was closer than was even slightly comfortable. . . .

The weather is still—Kansas. I've been here a month and flown eight times; at that rate, I could spend the rest of the year here, happy thought. . . .

March 8, 1944

Dear Folks,

I am enjoying a quiet evening in the music room, reading, writing, and at present listening to Tibbett sing "Wotan's Farewell." The last two days have been wonderful. Yesterday I had my first B Stage hop. As there was a 40K wind blowing, I had a Safety Pilot, which meant a dual instead of the usual solo. We shot a few landings and came in early to avoid freezing our faces. McDuffie (my instructor) was in a fine mood as he just that day made Lt. (J.G.). Today it was equally cold, but little wind. We did about half of the B Syllabus including S turns to circles and small field forced landing procedure. The main thing in B stage is precision in landing—a rather important factor in naval flying. Then third period I soloed. It gives me quite a thrill to taxi out, take-off, practice an hour and come back in, making a good landing. It's more than just soloing a new plane, even when that plane is about as sweet a job as I'll ever hope to see. I get quite a kick and no little feeling of pride out of taking up a plane with U.S. Navy insignia, giving the mechs the sign, and holding my own in Navy traffic. Also I feel pretty good about hitting two out of four circle shots twice the first day. That's as good as you're expected to do on the B check. I may have mentioned that nothing is so depressing as a bad ride; well, nothing is so uplifting as a good one, either. So that you'll know just what's going on, here are diagrams of these two maneuvers:—

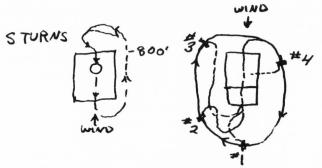

SMALL FIELD: flying a 1000' circle, the throttle may be closed at any of the points indicated. A landing to be made with a slip on first ⅓ of field. #4 is at 200' on take-off & is cross wind.

This is the kind of flying I like (not yet having had acrobatics) and it's something that demands plenty of concentration, work—and luck. Yesterday was a big day for minor crack-ups. One fellow nosed over onto his back right in front of the hangar. Many wing-tips, but more embarrassment than damage in all cases. . . .

March 14, 1944

Dear Folks,

Just finished the box of candy, so I'll drop you a line of thanks and what news there is. We've had foul weather since first flight Monday and so haven't accomplished much. The 13th was an unlucky day for a couple of cadets who collided near an outlying field and spun in. One was killed and the other seriously injured. The one plane was pretty well demolished, but the other came in on one wing and this saved the pilot. Proves the old adage "Better a stiff neck from looking around than a broken one from failing to do so." . . .

March 19, 1944

Dear Folks,

Methinks I haven't written since the first of the week, but there really hasn't been much to write about. Monday we flew, then three days of bad weather. Friday we flew, and now two more days of bad weather. Definitely not the sort of weather I'd ask for. The bad weather gets on everyone's nerves and you can't learn much, or very fast, flying twice a week.

To liven things up, the Welfare Dept. threw a big stag Friday evening after the show. Everything was free, food and drink, and the ham sandwiches were made of real ham. We all turned out—strictly cadet—and had an uproarious time, singing and being entertained by the local talent. One feature was that we had taps at 2300, but we paid for it the next day.

Among the various films we saw in the numerous foul weather programs this past week was a group of British pictures of various Allied and German planes in action. It was most interesting and gives you an idea how the planes look in flight—quite different than at $\frac{1}{50}$ of a second on a still slide.

Having neither liberty nor flight this morning, I finally managed to get to church. The service was very nice, but far from that at Iowa Preflight. They have no choir, but are trying to organize one among the tarmacs. . . .

My week of evening navigation classes finally dragged to an end. I made a 3.9 on the make-up test to redeem myself. Also raised my average in Recognition, Aerology, and Gunnery this week. . . .

March 20, 1944

Dear Phyl,

Perhaps it's just as well that I didn't flood the mails last week, as it was one H of a week. Every possible type of bad weather and in such quantities that everyone was in a foul mood. Friday it cleared up and we ran around joyfully in shirt-sleeves, but Saturday night we had a sleet storm. Today we are flying, but it looks like more weather is on its way. . . .

Anyway—Friday night, as everyone was so down in the mouth, the Welfare Dept. threw a big stag after the show. All the free beer and sandwiches you wanted, and plenty of rough and ready entertainment. I did well by the sandwiches, but I fear I'll never be a great tippler of the hops.

March 22, 1944

Dear Folks,

. . . This has been a good week so far. We flew Monday, and thereby hangs a tale—and also one wing-tip to my credit. I got it on the main base, a very minor rip in the skin of one aileron, but making necessary a damage report, none the less. Today we were forced to make all landings crosswind and I made a perfect one this time, so the mishap had its good points after all. It's a common thing here. We have a club of W.T.'ers in our platoon now numbering seven men, so you see I'm not alone.

Today we had a low ceiling with a few holes and a shifting wind, so finding low work impossible and high work alright on my part, my instructor took us up to 7,000 feet to look at the top of the overcast. It was beautiful. White cumulus clouds reaching as far as you could see in every direction, topped with a dome of turquoise blue that was as clear as a vacuum. The cloud tops rolled away in ridges and gullies below us and must have looked much like the widest and whitest stretches of the Antarctic. It was one of those things that makes you glad to be alive and be flying—and know that you'll keep on flying in the future, whether for the Navy or not. Sometimes when we haven't flown for two or three days at a stretch, you don't care whether you fly again or not, but once you get up—well, there's nothing so apt as the trite phrase—"The sky's the limit."

Today's flights took me half way through B Stage and I've plenty to learn in the second half. I'm hoping we'll have some decent weather for a while now, so that I can get some good work done.

Biggest achievement this week was made today when I shot an 81 on the pistol range. Four bull's-eyes and four 9's and an 8 looked pretty good. Latest development in Navigation is determining sunrise, sunset, and length of twilight for any given position and date. In Communications we are beginning on voice procedure in radio-telephony.

Saw some more old faces today when the new group came in, fellows from the 6th Batt at NFPS I hadn't seen since last August. Also a number of fellows from Lawrence W.T.S. Four of my old Liberty roommates are together again up at Grosse Ile NAS, Michigan. The only good thing about it seems to be its proximity to Detroit. They all wish they were in sunny (???) Kansas. How much rain have we had already this month?

. . .

March 23, 1944

Dear Phyl,

... This has been a typical week—bad weather, although we have flown, up 'til today. Now it's beautiful out, and I have instructor's liberty. Yesterday we had a cumulus overcast, and finding practice impossible because of crowded fields and low ceilings, my instructor took us up to 8000' to look at the top of the cloud floor. It was magnificent. Pure white hills and valleys rolling away as far as you could see, and above and around a pure turquoise sky. It's one of the things that makes flying worth any work or disappointment, and something the war won't end for me.

Lots of excitement at present over a PBY Catalina Flying Boat that just landed (I am not kidding) and is drawing a great crowd of onlookers.

About the number of letters I receive from you, Phyl. I certainly do *not* mind and would be very sorry to have the influx cease. I am very unpopular because of the amount of mail I get regularly, but it would be disappointing to at least twenty cadets who enjoy swearing at my pile of correspondence, if it should be decreased by you. Incidentally, I'd kind of care, too. The cross-word puzzle almost got me down. I sweated over it for five days before light broke, and lo and behold I had almost all of the right answers. It would serve you right if I'd send you one full of Navy slang, but time won't allow, so I'll be kind and let you escape.

March 27, 1944

Dear Folks,

Just a note to supplement Sunday's call. Saturday afternoon we sat around for three hours waiting for the ceiling to lift and then flew one very cold period. Getting down at 1745 and catching a bus at 1815 meant a bit of a rush. Met T—— at 1945 and we had dinner and a swell visit. ... T—— likes his work, although [he] admits it's about as tough as it could come. Likes Lawrence about as much as I did. Much in doubt as to how long he can stay in school. It's all up to his draft board. He had to go back that night, and I had a good bed at the Pickwick.

More foul weather Sunday and so I went into Olathe to see what it was like. Didn't stay long, but saw all the town. They have a pretty decent USO for a little town, but that is absolutely all.

We flew one period today and then the ceiling came down and stopped all flying. I have ten more flights before B check and am not doing any too well, although not particularly worse than anyone else. My instructor isn't worried yet, so I guess I shouldn't be.

We finally got our travel money from Iowa, hence the enclosed money order to add to the pile. . . .

March 30, 1944

Dear Folks,

Half an hour to relax, so I'll drop a line home. This has been a hard day. Two flights and two tests. It was exceedingly cold this morning and we wore everything we could lay our hands on. Wednesday we spent an hour knocking snow and ice off the planes in order to fly one period. The rest of B stage will be flown two solos and one dual—for nine more hours. It's going to take plenty of concentration and plenty of luck. We have had the pleasure of seeing a C-54 Skymaster around the base the last couple of days. It is a beautiful plane—silver, four engines, and a wingspan of 118 feet. Biggest U.S. land-based operational plane and fully as impressive as the slightly larger B-19. . . .

Really isn't much to write about. We're having wrestling in P.T. at present. Fairly enjoyable. We have been issued new summer helmets— a sort of waterproofed, khaki, rayon-like cloth with new and excellent Gosport tubes.* Only trouble is there were only two sizes, large and extra large.

Recognition is getting pretty rough, and the Aerology test today was really a stinker. My marksmanship seems to be improving steadily. Latest twist in navigation is a method of more or less guaranteeing arrival at a point when you don't know where you are. Pretty good idea, eh? All you have to do is be able to see and shoot one celestial body, sun, moon, star or planet. The only trouble is that the more you know, the more errors you can make, and we haven't touched on the use of

---

* The Gosport was a simple flexible tube that provided minimal voice communication from the instructor to the student.

the instruments with which navigation is done. I don't believe I mentioned that my notebook average in Nav. for the first half of the course was 4.0. Only one in the class. However a couple of bad tests reduce this to a moderate and less ethereal average. Picked up 12 w.p.m. in code and 8 in blinker yesterday. Good enough to check. Shouldn't have too much trouble raising the code to 16 w.p.m. with 80 minutes more practice each week now that we have completed the course on communications procedure. That much can make plenty of difference. In a recent letter Bob P—— mentioned that he is already taking 16 and 10 in radio school now. However they probably have about 4 hours of code and blinker every day. . . .

April 2, 1944

Dear Phyl,

No, I haven't forgotten you, but I have been pretty busy flying this week. We've had a run of good weather and have been piling up the hours. I have two more hours until my B stage check. So far in our outfit the "ups" and "downs" have been about equal. I hope I can get a good flight and an "up" on the first ride. However, I can't see any good reason for being very optimistic. But that's the way it goes. If you don't get it the first time, you will the next time.

Had an enjoyable hour last night scanning a collection of Dorothy Parker. I was surprised by the contents. Also last night was the Cadet Regimental Dance. What a ghastly affair. I found myself attached to a high-school senior who was passably fair, but not an intellectual by any means. To say I was bored would be a kind understatement. The base band was the only good part of the evening. . . .

The entire life of the barracks at present is entirely hooked up around the coming check-rides, and it's hard to get anything like that out of your mind. Everyone is sitting around trying to write letters and thinking about those circle shots and how in the devil they are going to fly an up. And then some guy comes around and tells about how he hit all of his precision shots without a bit of trouble, or else how he almost got an "up" but not quite. So you'll pardon me if I sound a bit preoccupied—I am. Just hope that in a couple of days I can write a nice happy letter telling all about the aerobatics we're taking up in C stage.

April 3, 1944

Dear Folks,

Don't have much to say, but thought you might like the new letter-
head. The flying is going as well as I could hope for, and with a touch
of luck I think I can make the B check. The weather has been wonderful
here, and flying has been everything it could possibly be—for which
words are quite inadequate. I'm not sure when I'll get the check. Should
be tomorrow, but I think I'll have two days off on instructor's liberty
and then a last solo and check. Got a very low letter from P—— at
Grosse Ile. He and the rest of my friends and buddies up there are
having a bad time; R——, the boy from Hays [Kansas] who had the top
grades at Liberty has two downs and may be gone by now. Time passes
very rapidly now that we are flying three hours a day, and every minute
(except the one before reveille) is a pleasure. . . . My reading has turned
to a collection of the best American plays. I am currently getting a whale
of a laugh from "Life with Father." The time might be spent more
profitably studying, but the mental relaxation after and before flying is
worth more than an extra point in Aerology. Some K.C. drama school
put on a "musical comedy" Sunday night. Over half the audience walked
out, but I groaned and bore it. It smelled! But the people were honestly
trying to amuse us, so I felt I owed a small amount of respect. Ate dinner
at the canteen tonight for a change—chili for the first time in a *long*
time tasted pretty good, but on top of two milk shakes, cup cakes,
peanut butter sandwich, pickles, milk, popcorn, and coke—well, let
your imagination fill in. . . .

April 7, 1944

Dear Folks,

Everything is about as bright as the weather today. With yesterday's
check ride over—a thing of the past, I can concentrate on something
new with a fresh mind and outlook. I didn't let myself worry about the
check, but I was plenty nervous, believe me. In our platoon eight have
had B checks; four of us got ups and four got downs, so you can see
how it goes. Of the four who got downs, three have redeemed themselves
with two ups and I am sure the other will. Two fellows also had special
checks which they passed. My check pilot was a pretty good Joe (also

my instructor's best friend) and made the most of my good points and overlooked the bad ones. The fact that our airspeed indicators were not synchronized didn't help much as I had to do everything 5K faster than usual. This means a lot of difference in the air.

Anyway, today we started C stage aerobatics with slow rolls, snap rolls, and loops. The maneuvers are just what they say—a slow rolling about the longitudinal axis, a fast or snap roll which is one of the most violent maneuvers, but also one of the easiest, and a simple vertical loop. It is the easiest and the nicest. A slow roll is pretty tough as it takes a lot of cross-coordination, strong handling of controls, and most of the time you are hanging by your safety belt. The cartoon which shows Dilbert hanging from his belt like this: is not quite the way we do it. All work is done above three thousand one hundred feet indicated (two thousand actual) so even if you stall out and spin, there is time enough to recover a number of times. I really like the aerobatics thus far. Inverted flying is really great once you get over the fear of falling and relax and start looking around. A well executed loop is one of the smoothest and most beautiful maneuvers you can imagine. Only trouble is that a plane won't fly itself through aerobatics and you get a pretty good workout in two hours of the stuff. Pity the poor instructor who rides all day!

Too bad about B—— W——. That's the trouble with this racket. You spend ten or twenty months training, get to the point where you can take care of yourself in the air, and wham, you're not in the air anymore. One of our instructors was killed yesterday in a crash. Cause unknown. The cadet apparently will survive. Evidently they were caught with insufficient altitude and had to ride it in. Basic cause—pilot error where there's no room for error. . . .

April 8, 1944

Dear Phyl,

While you are enjoying the Big Town, I am jubilantly kicking my heels, whistling, and singing my great joy at having checked out of B stage. The check ride was not too good, the check pilot was definitely a Santa Claus, but the fact remains, I got an up, which erases all cares and worries from the face of the earth. Tomorrow I start in on acrobatics. We get twenty hops and then another check. Today's wind of about 25K didn't help matters much. Served me right for taking instructor's liberty the last two days when the weather was perfect. It looks like we'll have a spot of precipitation some time tonight. . . .

April 9, 1944

Dear Folks,

It turns out to be a beautiful Easter despite rain and clouds in the morning. It was a perfect afternoon to fly. Little wind, beautiful clouds, unlimited visibility, and enough holes in the clouds to get in acrobatics. I'm really enjoying my flying now, although it is plenty tough to sit through three hours of acrobatics. I've had the whole works now and it remains only to perfect them. I mentioned the loop, slow roll, and snap roll. The other maneuvers are Immelman, split "S", "cartwheel," and "falling leaf." The Immelman goes like this:

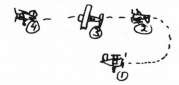

and the "split S" is just the reverse:

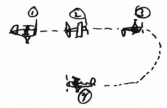

The "falling leaf" is a stall maneuver in which the plane falls in the same manner as a leaf, with oscillations. I'm coming along pretty well I think. At least my instructor seems happy, which is a good sign. Today we worked up over toward Lawrence, and when I told him I went to WTS there he went down and we landed and looked the place over. No one around, but the old Cubs and N-3-N's are a little more beat up than before. It gave me quite a thrill to remember how I had felt last fall seeing the fellows from Olathe come over doing slow rolls and slip landings. The attrition (washout) rate has jumped up considerably in the last couple of days. I'm very glad I have the B check behind me. Today marked the first quarter of C Stage. I hope to get my acrobatics worked out by the halfway point and really polish things up for the C check. After that, if successful, comes the front cockpit work on formation and night flying in D stage, and then I'll be all set for 18 days leave, I hope. . . .

A couple of other maneuvers I've had are the inverted spin, in which you are on the outside of the rotation—really throws you out against the belt; and a reverse Immelman in which you go through half an outside loop and then roll over—it was pretty strenuous and not part of the regular syllabus by any means.

Prize laugh of the week came over the PA system yesterday afternoon when we heard the announcement, "Cadet ———, your father is in the brig." You see, the brig also serves as a waiting room for civilian guests.

All of which covers the subject pretty well. I would appreciate it if you would hunt up a pair of my unlined gloves and send them to me. Lined mittens are too warm and sticky these days. Today I flew with nothing more than my regular leather jacket and was very comfortable. . . .

April 10, 1944

Dear Phyl,

By this time you should be back at your work and I will be well on in C stage. Thursday's highly successful check ride left me in rare high spirits, and the progress of events since that time has maintained this level. My new work in C stage consists of a number of acrobatics such as slow rolls, snap rolls, loops, Immelmans, Split S, Cartwheel, falling leaf, inverted spins. It is very interesting and strenuous work which, I am happy to say, I am getting on with in good fashion. Both my instructor and I are quite pleased with my progress so far in this stage, five of whose twenty rides are now past. We have had, until today, a wonderful run of perfect spring days in which flying is a real pleasure. Saturday night we had a thunderstorm which broke up about 1000 Sunday morning and gave us a perfect Easter Day with great buttes of cumulus jutting up to 7000 feet, and hardly a trace of wind on the ground. After an hour and a half solo dodging around, into and out of cloud banks trying to find space for acrobatics, I had a dual period. McDuffie (instructor) was in rare good humor and we climbed up in a northwesterly direction doing our stunts. At the end of forty minutes we were over the Lawrence field, and when I told him that it was my WTS alma mater, we went down for a smoke. The place has been greatly improved, but I'd hate to have to fly those planes again. It gave me quite a thrill, remembering how I had felt last fall about Primary, Stearmans, and Olathe. I chuckle at the picture of the chief pilot at WTS showing me a snap roll which was no better at best than those I am doing now. It seems I was rather fortunate in getting an up on my first ride. In our platoon, which has been more fortunate than some, we have had a ratio of two downs to each up check. Half of the former group redeem themselves as I had to do in A stage, but it isn't a pleasant or easy task. I should get my C stage check about the end of next week. More than ever it's got to be an up so I can get on to front cockpit, night flight, and formation. It looks like I am at last getting someplace—the light of progress breaks through the overcast after ten months.

We enjoyed a very nice quiet communion service Easter Sunday which was accentuated by the contrast of the preceding night in Kansas City. I pretty much cured myself of heavy drinking again, and as I had seventy-five minutes sleep on the YMCA floor, the mental humbling of the sacrament fitted in nicely with my physical state. . . .

This letter has been interrupted by a call for all hands to the line on the double—tornado warning. An hour later all planes are secured, over two hundred in hangars and a number in the gymnasium. A severe tornado hit above Wichita, but word that it went through town proved false. They must have had quite a blow down there, though.

April 13, 1944

Dear Folks,

Your letters were very welcome yesterday. Although weather and flying and ground school all look pretty good, the morale around here is about as low as I've ever seen it. Yesterday a couple of admirals came aboard and before leaving suggested to the Captain that it was the new policy of NAPTC [Naval Air Primary Training Command] to eliminate as many cadets as possible. Specifically stating that outstanding work in all phases but one would not make up for any single phase. In other words, the Navy has cut its pilot requirements for the next three years and so is proceeding to get rid of the surplus as fast as possible. Down-checks are coming thick and fast. And everyone is pretty well disgusted with the turn of events. However, we'll keep on doing our best until the end comes. . . .

The night of the big storm we had tornado warning here and everyone was out on the line moving and securing planes. They had about two hundred in two hangars, on their noses, and even in the gym. Heard rumors that Wichita had quite a blow, but got the word from Aerology that it wasn't serious. . . .

April 14, 1944

Dear Phyl,

From the contents of your last letter I gathered that the New York jaunt was a success. . . . Had to laugh at your experiences with genus homo intoxicatus. They, or should I say he, can be most interesting, amusing, or disgusting. The type I personally prefer is the service man on leave who in other circumstances is a pretty swell Joe. Met a typical one last weekend, an army captain back from Guadalcanal and Rendova. He was O.K. and nothing was too good for the Naval Air Corps. They were the boys who saved his skin in the Pacific, and we had a great time listening to and drinking with and on him. I feel sorry for the male civilians when a bunch of service men get together. They have more fun than anyone.

I am having a swell time in C stage and coming along well enough, I believe. I have ten more rides before I check and haven't begun to worry yet. However, a couple of admirals came aboard this week and passed the word that Naval Aviation is pretty well filled up for the next three years, and so if the wash-out rate rose to about 50% the Navy might profit thereby—and save a lot of money. Consequently the downs are coming thick and fast, and the whole barracks cheers when someone comes back with an up. Having passed the usually crucial B check, I just can't see washing out and have every intention of sticking as far as Corpus Christi anyway. Also there's a nice fat leave after Primary, too, that beats the five days you get if you wash out.

News of anything except flying I have little of. . . . Your last letter just arrived and considerably brightened up an otherwise gloomy day.

We tried to fly with a 2000′ ceiling this morning and so most of us in C stage "incompleted" the hop; most acrobatics begin twice that high and end there.

April 17, 1944

Dear Folks,

Another Monday afternoon and bad weather. However, as I have instructor's liberty and can't fly anyway, I don't mind a bit. Spent the afternoon profitably—clean laundry, haircut, dental appointment, some reading, navigation, and a letter home before dinner, so you see I don't waste time when I have it. . . .

I am half thru C stage and should check Sunday morning if we fly the rest of the week. I feel like I've gotten some place; if I can pass this check the end of primary will be in sight—and that means a lot from where I sit.

Pretty good quantity of mail today—although none from home. . . . My friend P—— at Grosse Ile is having trouble in B stage and is on extra time.

Main news item is my recent weekend, I guess. I got overnight liberty Saturday until Sunday night; however I had to stay on the base 'til 1800. At inspection Saturday we broke out in white tops—looks pretty flashy after a winter of solid blue uniform. But it's not so nice trying to keep them clean. I hate to think what it will be like if and when we put on whites. Some of us have them, others only khaki, so I don't know what the summer uniform will be. Probably stick to blues with white tops except for liberty.

To get back to the weekend. I got into K.C. about 1915 and had dinner with a couple of optical students from Chicago down for examinations. Took the 2135 Rock Island to Lawrence and arrived at 2300. Went right out to T——'s room and we talked a couple of hours and so to bed.

I told you that he is rooming with our WTS P.T. instructor. His wife is organist at the Christian Church. Another pre-med stays there—a nice fellow who served a stretch in the army and was discharged. The house and room is very nice—he couldn't ask for anything better. Lots of space, windows, private bath, good bed, two dressers, desk, one block from the Hill, and a musical household. The other fellow is first hornist in the KU orchestra. Needless to say, sleeping until 0830 was luxury. Breakfast at a local hangout was very good, and then to church. T—— is bass soloist at the First M.E. of course. It is a beautiful little church with good attendance, an amazingly good choir, and a fair preacher.

We had dinner at the Colonial Inn, Lawrence's best dining room. Had a wonderful steak dinner. Met Charley B—— on the way back and had a short visit. Also saw M—— C—— and L—— J—— who were at F.U. last year. Spent the afternoon until 1530 in a soft chair enjoying the NY Philharmonic as I would if I were home. Caught a 1600 bus to K.C. and was back at the base at 2130. Had dinner with an old fellow who claimed to be a big meat packer and invited me to visit his plant. You meet a lot of interesting people at dinner time. Well, it was a perfect weekend, and I not only had a good time but got a good rest as well.

One of the best stories I've heard lately was about an R.A.F. pilot in the Mediterranean area, out on patrol one day he spotted the periscope of an Italian sub. Diving low, he daubed the periscope with green paint. The sub rose a little higher and a little higher, unable to see where he was, and when they reached 3000′ the RAF pilot shot him down!

This letter was delayed some twelve hours as we were called out on the line again last night to secure all planes and get as many as possible into hangars. A double hurricane warning was flown and 70K winds expected. As any Kansan would have predicted, the wind went down almost immediately, and today it's warmer although still overcast. . . .

April 21, 1944

Dear Folks,

. . . Yesterday I got up for the first time in a week—and boy did I stink! It was no day to be up. One hole in the overcast and a dozen planes doing acrobatics over it. There were at least five forced landings yesterday and one fatal collision—B stagers—we have been very lucky in our class—no one has gotten hurt although there have been some close calls. Today we flew with a complete overcast ranging from 3600–5400′ IND[icated altitude] (that is 25–4300 actual) and visibility was like a London fog—thick. I had an interesting experience when the right flap of my engine cowling came loose as I was diving to begin an Immelman. It raised quite a rumpus until I slowed up. In the dive it sang like the struts do in a power dive. Needless to say, I high-tailed it back to the station and landed before it came back and parked in my lap or bounced off my head. Consequently I've logged one scheduled flight in the past seven days. Not the way I like to do my flying.

Finals next week in ground school and then we'll be hold-overs. Got 16 and 9 in code and blinker last week—hope I can do that on the final. . . .

April 21, 1944

Dear Phyl,
. . . Yesterday we flew again after a week's lay-off. One hole in the overcast and a dozen planes in it makes for fun. There were at least five forced landings yesterday, and a dozen wing-tips today—you can guess what we think of the weather. . . . I had a little fun today. I was playing around in the soup around 5000′ doing acrobatics. Dove down out of a Split S for an Immelman and noticed an awful funny whine; as nothing let go up to 125K, I pulled up to an angle of about 90° (vertical) and lo, and behold, there was the left side of my engine cowling resting against the flying wires. Believe me I let it down easy and headed for home—I don't feature having a piece of tin bounce off my head or come through my windshield.

A couple of boys collided yesterday as per usual when we are rusty and the weather bad. One hit the silk and is still with us. The other spun in and burned. Of interest was the fact that they identified the body by some recent fillings in his teeth. Hm.

April 26, 1944

Dear Phyl,
. . . So—the humidity in Kansas has been rather high lately, if you've been reading the local papers. Wichita has been entertaining Little River and the Canal all over town, including 18 inches in Dad's warehouse. Quite a mess, I guess. In the middle of which, I went hightailing it to Lawrence, catching the last train in (Santa Fe) and the last train out (Union Pacific). With Santa Fe tracks and highway under 6 feet of water, I got up at 0530 Sunday morning to watch five trains go toward K.C.

without stopping. I finally got one and just made the last bus out to the base.

Waiting in the K.C. Union Station Saturday, I saw Bob H—— who was quite jubilant over a very fresh commission in the U.S.N.R. All you have to do nowadays to get a stripe is play football or coach athletics. Nuts, phooey, and even a mild swear word. Nice for Bob, though, who looks the same as ever.

I haven't been up for six days, and we have clouds at 50′ at present, so it looks like I'll be here 'til the base closes in June. Fine stuff. We're having ground school finals at present, so when the weather breaks I should get in 3 flights a day and finish with a bang (or a pop!). Everybody, including me, is about ready to move on to another base.

April 30, 1944

Dear Phyl,

Here it is Sunday morning and we still have a nice low overcast of clouds of the type known to cadets as "secure-us clouds of foul weather." And here's one cadet who wants to fly and fly high and fast.

Yesterday we took our last final, winding up ground school in its regular session. I am satisfied with the results, in view of the little amount of work required. I pulled out a 3.8 in Navigation, 3.7 in Recognition, 3.6 in Aerology, 16 w.p.m. in code, 8 w.p.m. in blinker. All very satisfactory. For the remainder of our stay we'll have one Navigation class, three communications classes and P.T. four times per week. That's a pretty soft schedule, which I hope will allow us more flying time. Yesterday I was pretty hot on all my work, and I'd sure like to carry it right through and check C stage. I'm not worried about the next check despite the fact that it now seems to be standard to get a down on the first check and then sweat through two or more up-checks. . . .

Monday evening:—The foul weather came thru and we were off on liberty from 0900 until 2200. I spent the usual dull day in K.C. seeing two lousy shows and a worse stage show. The penalty one pays for the company of others is the sacrifice of his personal—and private— pleasures.

This has been a perfect spring day, until this evening. Warm enough to discard coats and perfect temperatures for flying. However, I suffered a "relapse" for my Sunday vacation. We knocked off at 1630 to offer up a prayer for the soul of the late and lamented Frank Knox, Secy. Navy, which took an hour of standing in ranks and sweating, and also robbed us of a flight. My next one is #44, check on #50, so I'm still determined to check this week, despite the present outlook.

The rumor that Olathe will close as part of NAPTC in June is pretty well founded, and we are all praying for the necessary good weather to get us out of here before then. Two weeks will do it for me, but at the present rate, it will take two months.

May 1, 1944

Dear Folks,

May came in as it should—a warm day with no extra gear needed for flying. However this evening the clouds came up again and there is no night flying.

Flight secured Sunday morning and we went into K.C. for a rather dull afternoon. One flight this afternoon, and another cancelled because of memorial services for the late Frank Knox. The C.O. appeared and read a few words to the assembled and sweating personnel—cadet regiment, officers, and a good part of the enlisted men and Waves. All very official, but not necessary.

If all goes well this week, I should check C stage on Friday. Of course I've been saying that for the last two weeks. There is a very slight chance that I might get home this coming weekend, but I'm not planning on it, and if I do check C, I'm sure it will not work out. D stage liberty is set up differently and is as yet a mystery to me. However, I'd be quite willing to sacrifice a weekend in order to get a leave sooner.

Flight goes on well enough. Another crash yesterday—an enlisted man and an instructor. My instructor is being transferred within a month, so I'm doubly anxious to finish flight soon. . . .

May 9, 1944

Dear Folks,

Another weekend and a couple of days out of another week—and no news. Yesterday we had low clouds, today a heavy ground fog and low clouds, so I still have two more flights before the C check. The trip back to K.C. was not very pleasant. Despite my getting on the train early, there were no seats, and so I stood all the way to K.C. A pleasant surprise, though, was to find Marvin S—— and his wife at the station at Newton. I had time to say "hello and how are you" before the train pulled out. Marvin has applied for a naval commission and will either get it or be drafted into the Navy within a month, so I was fortunate to see him at this time. . . .

Everyone is pretty fed up with not flying, and just lying around reading, sleeping, playing cards, chess, or writing letters.

Latest scuttlebutt is that the base will not be closed in June and a new draft will come next week. Hope to have more to write about next time.

May 10, 1944

Dear Phyl,

Tomorrow is the big day, I hope, and I check C stage. Had my last solo today and if all goes well I'll be flying formation by the end of the week. I am somewhat confident, but I've got my fingers crossed five or six times. Three of our boys checked satisfactorily and one got a down, so the breaks will make a lot of difference. . . .

To all appearances, this base will not close in June. Latest scuttlebutt has it that a new draft will come in next week. Not so good is the threat of discontinuance of completion-of-training leaves. At present the boys are getting 15 days with possibility of cancellation without notice—one fellow got 45 minutes at home and was recalled. Maybe by the time I finish there will be a great enough back-log to provide a good leave again.

This sitting around with no ground school to take up time is getting me down. I've turned to cards to while away some of the time. You'll be happy to know I've picked up the rudiments of bridge and double solitaire. My letter writing has suffered, but even I can't write everyone when there's nothing to say—which is what there is to say at present. It's still pretty cool here. Last week, May 5, to be exact, we flew in a snowstorm. We keep thinking that the sun will come out and the temperature jump 20°, but it just doesn't. It's clouding up again right now.

We're having a big dance out here at the base Sat. night and uniform will be optional blues or whites, so I guess I'll get to wear the whites at least once. I hope it will be a lot more often, though.

May 11, 1944

Dear Folks,

Today was the big day, and I am now in D Stage, having successfully checked C on the first try. Needless to say I feel pretty good about it and no little relieved. I am the fourth man in our platoon to check successfully. One fellow has two downs and one up, is up for squadron time; another has one down and one up. I am hoping to get a formation with fellows in my platoon, but this seems unlikely at present. I had a pretty good check ride, with only one questionable maneuver. The acrobatics were all good, and my small field shots were all good, so the check pilot said it was a pretty fair up-check.

Tomorrow I'll get my front cockpit check-out, which is one dual with your instructor, and all flight from now on will be from the front seat. No more "backseat driving." Formation should start the next day after a ground period on formation flight. Night flight is filled for a long time in advance, so I'll probably finish day flight before I get the night flight.

As I see it, there shouldn't be any trouble finishing here now. The D check #72 is about the same as the C check except it's flown from the front cockpit, and no one has ever washed here in Form[ation] or night flight. So, I feel a bit elated with everything.

Summer is here, if sunburn and sweat are any sign. Comes a sunny day and a hundred cadets tear out to expose lily-white skins to ol' Sol and they turn a beautiful pink. Monday we are changing to khakis and black shoes, which will be a change for the best. . . .

Saturday we are having a station party and whites are optional, so if it's at all nice, they will be in order. I exchanged the white shoes I got at Iowa for a pair that fit, so I'm all set. I've often wondered if I'd ever be wearing them. I guess the time has come at last. . . .

Don't count too much on my leave. A number of fellows came back today who just made it home when they received word that their leaves had been cancelled, and they are heading south in a day or two. In order to hurry things along, all extra-time hops have been cancelled. This will cut out a lot of flights that were very welcome when we were grounded for any length of time. . . .

May 15, 1944

Dear Folks,

It seems like a long time since I last wore these khakis—and today it feels pretty good to get into them. Yesterday I flew in my shirt-sleeves and was hot enough to be ready to take off the shirt.

This formation flying is really good—and you have to be good to do it well. I've had two of the nine flights, and at present we are flying a Vee formation which looks like this:
The difficulty lies in maintaining your  position at all times, particularly in turns. The other two fellows aren't so hot, and so we have a kind of rough formation, but I have hopes that time will iron out the difficulties. We ride with different instructors each time, so you get a variety of ideas on how to fly it. On my second ride the instructor said I flew a good lead and No. 2 position (left wingman), so I think I'll have no great difficulty. This is all flown from the front cockpit. The wingmen have to watch the lead man constantly, so his job is to keep a good lookout for other planes and to fly as smoothly as possible. The most difficult thing we've found is making a formation take-off. The lead man either gives throttle too slowly and the wingmen get ahead, or he runs away from them before they can get off the ground. I think we can work this out today, though, when we have our first solo hop. . . .

May 16, 1944

Dear Phyl,

Boy, oh boy, am I having fun now! This formation flying is really great. It beats anything yet. Today we got a solo hop on vee, echelon, crossover, change of lead, and break-up. Hard work, but the most fun flying I've found yet. Our nine flights are going rapidly, and we will probably check day after tomorrow. I feel pretty sure we can make the grade alright as long as no one fouls up some maneuver. After that it won't take long, and night flight is rapidly returning to normal, which means another two weeks at most. (Great cheering, here.)

We are enjoying Kansas summer at present. Most of the fellows are blushing beauties after a couple of hours in the sun. Very amusing to a native like myself who isn't greatly affected by the sun after so many years.

The days pass slowly. With two Form[ation] hops in the afternoon, the morning seems a separate day. The only trouble with D stage is the lack of liberty. From the looks of things I will be ineligible for liberty because of night flight just as soon as I get to the point where I can ask for it. Fine thing.

I appreciated your last letter concerning your forthcoming visit and am certainly looking forward to it as much as you are. Leave is still an unknown quantity. At present they are giving two weeks subject to immediate recall, so no telling what it will be if and when I reach that point. They are washing the boys out fast and furious. Extra time is practically a thing of the past and the only way to get an up is to fly one.

We still haven't gone into whites yet, and last Sunday blues were about tops in discomfort. Sure seems good to fly in khaki again and in your shirt sleeves.

I fear I am too wound up in my flying right now to concentrate on letter writing, even to you (you can imagine what my others are like—ye gods!). Some of the fellows are pretty luke-warm about it, but here's one cadet who has never found anything to equal it. . . .

May 19, 1944

Dear Folks,

Well, the show at Olathe is about over. Yesterday we checked out of formation with a good strong up. It was really fun. The last two days we made formation landings in addition to Vee and echelon—that was the best yet. If I shouldn't get on to Intermediate, the thing I'll be sorriest to leave will be Formation. That is really flying—military flying.

A couple of days ago we got our ground period on night flight and should begin tonight or tomorrow night. Consequently no more liberty for this boy until I leave. Which event isn't far off. I have six flights to get my work down pat from the front seat and then the final check. So if the weather stays as it has been the last couple of weeks, I'll finish day flight Monday, and night flight only takes three nights, so I should finish everything by Wednesday. Of course there's always the chance that all won't go well. Yesterday five fellows checked 72X and four of them got downs. I'm hoping this trend will change by the time I check. . . .

May 22, 1944

Dear Folks,

Another week begins none too auspiciously. Today was the first day in three I had flown. Friday my instructor had liberty, Saturday the weather was bad, Sunday we were scheduled for two night flights and so no day flights, and then they secured night flight. Under the new setup you can only fly two periods per day, so that will lengthen the rest of my flight. The earliest I can possibly get out will be Saturday, and it's quite possible Sunday or Monday. I'll get home the evening following the day I finish flying. This isn't the way I'd like to have it, but there's not much I can do about it except hope for good weather and fly an up on the first ride. P—— checked 72X successfully today and will be heading for Wichita in a couple of days as he already has two night flights. . . .

May 22, 1944

Dear Phyl,

As you see I'm still here and from the looks of things will be here another week. Thanks to a combination of unfortunate events, I had a three-day vacation from flying and so fell behind. At the same time a new ruling was passed to the effect that we can only fly twice daily, so here I sit with ten more flights to go. This means I won't get away before Sunday or Monday of next week and will probably mean a very short leave. Fine stuff. I still haven't gotten a night flight in, so I can't tell how that's going to turn out. I have one dual and three solos left in day flight before the final check. Then five night flights and I'm thru. This recent three-day lapse hurt my flying somewhat, and I'll have to work hard to get up to the desired point by Thursday when I may check.

May 25, 1944

Dear Folks,

The gods were smiling this morning and I flew my last solo and my 72 check with an up. This last check was a rather unpleasant and trying experience. Lasted two full hours, and by the end of that time I was ready to fold up. Parts of the flight were good and parts stinky, but as long as it's an up. I shouldn't kick.

So that's the end of day flight. Now for five night flights, and I'm ready to pull out. If the weather stays good I'll fly the last night flight Sunday evening and be on my way home before noon Monday. . . .

May 25, 1944

Dear Phyl,

One more letter from Olathe—and this is the last. This morning I flew my last solo and the final check. I had a good boy to check with, and so I got the big fat "up." It was a sort of stinky ride and I played Dilbert a couple of times, but as nothing was definitely bad, he was good to me. I was really disgusted with the ride I gave him, as I'd been flying well in the previous flights, but then any "up" is a good "up."

Now I have five night flights to get out of the road, one dual and four solos. If all goes well, I'll be through Sunday and on my way home Monday morning. I fully expect at least one day's delay, but I'm hoping for the best. The leave situation is very much up in the air and there's no way of telling what will develop, but I should either get a good long one, or a damned short one. Just wait and see, I guess.

*Night flying was completed in due course, and Cadet Rea was granted a leave which he spent in Wichita. As Phyllis Edwards was also in Wichita on vacation, their correspondence suffered a happy hiatus; their relationship deepened to what might be termed an unofficial engagement. Rea returned to Olathe about June 15, although he would not receive orders for the next stage of training at Corpus Christi, Texas, for some days.*

June 15, 1944

Dear Phyl,

. . . I am not going South this week, if all goes as it's supposed to. I am only a supernumerary and not one of the draft list. This means I sit around here until Saturday, checking out gear and getting papers signed, and then about 0830 Saturday morning tear off for home again until Wednesday night. I don't relish the two days here, but four more days at home sounds wonderful. Whadaya think? I had hoped to turn right around and go home today, but this is the next best thing. . . . See you Saturday.

June 22, 1944

Dear Folks,

Back at it, although Olathe isn't the place I left, very few cadets around, washouts all over the place despite last week's 50-man Great Lakes draft.

There are forty of us going down to Corpus Christi Saturday. Four from my platoon and half a dozen others whom I know, one from Lawrence W.T.S. days. The word from the boys already at Corpus is that the washout rate has been officially hiked to 50%, and we used to consider 15% high. Oh, well, all the more reason to work hard, although right now I'm about as un-eager as it's possible to be. . . .

June 22, 1944

Dear Phyl,

My desk is my knee and your writing pad. For company I have a row of empty bunks and a handful of new cadets. My platoon just isn't here any more. Three are still flying, one waiting to go north, and four of us going south Saturday. This draft is large, forty of us, and is, of course, headed for Corpus Christi. I know a number of the fellows beside those from my platoon, so we should have a pretty good time.

I certainly can't get eager about it, though. I had just enough leave to dampen any enthusiasm I might have. Just hope I regain it when I get into those Vultees. Everything we hear here about conditions South is discouraging. Washout rate has officially been raised to 50%, which is terrific. But enough of this, can't expect to be eager immediately after leaving home and you.

June 23, 1944

Dear Phyl,

So this is the last letter from Olathe. I'll be pretty glad to get away, too, and on to something active and new. No doubt in a couple of weeks I'll be wishing I had a little of this leisure time to enjoy. That's life. Being already checked out, I've done nothing the past two days except rest, read, and listen to some new records in the library. Presumably because of the low morale on the base, an informal dance is being held every evening at present. Dropped in last night for cake and coke. Every once in a while the Navy forgets itself and does something very nice.

We are getting the fickle finger of fate tomorrow morning and catching an early train in K.C. So early is it, that reveille will be at 0345! We are supposed to have Pullmans, and I've a hunch we'll bunk down as soon as we hit the train.

I completed checking out this morning by getting my log book. It seems that I picked up a mere 107.5 hours here. Highest number of hours was 122.5, the result of a surplus number of check rides. That kind of time I don't mind missing.

. . . From the looks of things, it's going to get pretty damp around old USNAS before very long. Yesterday was hot as a firecracker and any change will be for the best. We are going South in khaki but will have to report at the base in blues. Here's hoping Texas is cooler than we anticipate.

# 6

# Winning the Wings of War

JUST AS NOVEMBER 1942, when Bob Rea joined up, was a pivotal month in the war, so was June 1944. At long last came the cross-Channel invasion, with its armada of warships and landing craft and an umbrella of virtually unopposed warplanes. The scenes that soon filled newspapers and magazines and moved across the screens in successive reels of Movietone News were some of the most memorable of twentieth-century warfare: the ramp of an LCI dropping, assault troops plunging into waist-deep water, the beach and distant cliffs wreathed in smoke; Norman fishermen in hipboots gazing pensively at the shrouded dead on the beach—their protruding combat boots a last token of individuality.

Other events in distant war theaters were equally portentous. The amphibious juggernaut in the Central Pacific struck the island of Saipan in the Marianas, while a naval battle three or four times the size of Midway raged in the Philippine Sea. The battle was inaugurated by duels across the sky between carrier planes. Nonflying personnel aboard U.S. carriers sat on empty flight decks, watched swirling contrails, and cheered like fans at a sporting event. In one day there were over 400 occasions to cheer in what came to be known as "The Great Marianas Turkey Shoot." In this the greatest carrier battle of the war, a third of the enemy's remaining flattops went down. The victory ensured the success of the landing on Saipan. In July, Tinian was taken from the Japanese. It would play a special role in the war's final act. Warming up

[ 175 ]

for that drama, the Boeing B-29 Super Fortress flew its initial combat missions in June. The construction of American airfields in the Marianas would provide a direct skyway to Japan.[1]

Bob Rea was on leave in Wichita during much of that eventful June of 1944. When he returned to Olathe, late in the month, it was to prepare for a train trip south to Texas. At Corpus Christi the pace of training began to drag even as the war sped toward a climax. In the months ahead, the pictures on the movie screens piled one event upon another: Patton's Sherman tanks, on dusty French roads, passing junkyards of violence created when Allied fighter-bombers swooped down on retreating columns of panzers; hordes of Russians with vast amounts of armor pressing back the stubborn but thinning German ranks; GIs striding through rows of concrete antitank "Dragon's Teeth" on the way into Germany; and a dejected column of American POWs shuffling along the snow-encrusted shoulder of a road toward the rear, while a Tiger tank whose crewmen sported SS runes on their collars lumbered in the opposite direction.

Another sobering development appeared in 1944, one that reflected the war's fanaticisms: the kamikaze. In October, when American naval task forces maneuvered to support MacArthur's return to the Philippines, the Japanese high command, greatly outnumbered in ships and planes, unleashed "the Divine Wind." Specially indoctrinated Japanese pilots were sent off to dive their explosive-laden planes into ships of the U.S. fleet. This desperate appeal to the gods of war would cause many casualties and sink many ships before it spent itself in the summer of 1945.[2]

From June 1944 to January 1945, Rea's time and interest were principally consumed by the last phase of cadet training at the sprawling Naval Air facility centered at Corpus Christi, Texas. During those seven months he was shuttled between the Main Base and three outlying fields: Cuddihy; Chase Field, Beeville; and Kingsville—in all, eight removes. At Cuddihy he completed Basic indoctrination in heavy, all-metal aircraft, flying the lumbering old SNV Vultee. At Beeville he went "under the hood" in the instrument flight training program. At Kingsville, the Advanced stage of training was flown in the most popular training aircraft known to World War II pilots, the SNJ (to the Army, the AT-6 Texan). As a fillip, an extra few hours, designated Pre-operational, were added in the Navy workhorse, the SBD Dauntless dive-bomber, prior to commissioning.

Sporadically but certainly, the experience at Corpus focused upon flying and the job for which cadets were preparing themselves. All else seemed a waste of time, particularly the waiting periods, referred to as "pools," between one stage of flight training and the next. Texas offered few attractions: Corpus Christi little beyond the yacht basin and the towering Driscoll Hotel, a landmark for lost Latin American trainees who circled it at night while pleading for landing instructions from the nearby field—or so the story went. San Antonio was a distant liberty town. The happiest memory of Texas was created by a wonderful couple in Alice who adopted a pair of lonely cadets one weekend, dressed them in their sons' chaps, mounted them on cow ponies, and led them through the mesquite looking for Brahma bulls—an occasion that had nothing to do with Naval Aviation but said much about Americans at war in 1944.

Time passed slowly between flights, and there was ample opportunity to put the whole experience on paper. There was also an added incentive, for the letters between Phyl and Bob had sealed the firmest of bonds—at least until the Navy would allow them to marry. Phyl visited Corpus during her vacation, and Bob's commissioning leave took him to Portland, Maine, where they were married—by no means the only couple in this compact segment of naval society to commit themselves for the peaceful future.

The months at Corpus Christi epitomized the Naval Aviation training program. Cadets entered the program full of zeal, intent upon learning a wartime profession, and scorned everything that delayed, interrupted, or failed to contribute to that end. The Navy frequently seemed to belong in that category. It could turn boys into men and cadets into officers, and it could teach them to fly; but it failed to turn civilians into militarists, and the harder it tried, the more it failed. Perhaps that was one of the wars Americans won in the mid-twentieth century.

## Intermediate, Instrument, and Advanced Training, U.S. Naval Air Training Center, Corpus Christi, Texas, June 25, 1944–January 24, 1945

Somewhere in Mo.
June 24, 1944

Dear Phyl,

Now I know what the pioneers went thru riding covered wagons. It's a beautiful day, and we are passing thru a beautiful part of the Missouri Ozarks on the Southern Belle, fastest streamliner from KC to New Orleans; but despite the fact that we are traveling on first class tickets, we are stuck in a couple of troop sleepers that ride like box cars and are just as comfortable and warm. Hence the illegibility of this effort—it's worth my life and limb to try to write, the train jolts so I'm apt to stick myself in the leg. We just stopped at Gravette, Arkansas; do me a favor and look up a map and tell me how the dickens we are getting to Corpus by way of Shreveport, La. I can't figure it out. This is nice country to travel through, black-jack covered hills and a hundred little gravel-banked streams. Looks a little like southern Indiana. I'm giving this up for a while until we hit a little smoother track. . . .

[June 25] Sunday noon—Houston Texas: What a ride! Southern Oklahoma and Arkansas were beautiful—high green hills, pine-scented air, and light intermittent showers that cooled the sultry afternoon miraculously. We left the Southern Belle at Shreveport. It was far from the lovely southern town our Louisiana buddies had pictured. We were there an hour, and half the draft was drunk when we pulled out. The town was hot, dirty, and full of soldiers.

The night was hot and dirty, and we picked up several cubic inches of Texas red dirt to add to our discomfort. So naturally Houston and a shower were most welcome. It's a nice town, about like Wichita, but more so. At present we are leaving Houston for Corpus, due to arrive about 1845.

We've all been looking around as hard as possible for the beautiful Louisiana and Texas women, but none found as yet. Most of those we have seen are pretty sad sacks. Not a man among us (those of us who are Yankees, that is) but that would give his last cent to be North again.

Lots of reliable scuttlebutt running around. The papers today carried the official announcement that 50% of Navy aviation cadets are to be washed out, so everyone is in a pretty carefree mood. Some talk of leave between Basic and Advanced, too.

Corpus Christi<br>
June 27, 1944

Dear Folks,

I'll start this now before evening chow and finish whenever possible. We are well into indoctrination week and, for a change, not over-worked. We got summer helmets, goggles, and very nice light leather gloves and light flight jacket. We've had lots of time to rest up, sun-bathe, and swim. The pools here are outdoors, warm, and very nice.

The base is out on a peninsula with bay on both sides. PBY's and OS2U's are in the air all the time. I would guess that there are 200 big PBY's on the base. A few figures:—20,000 acres, 3,000 officers, 6,800 cadets.

We have nice outdoor swimming pools with warm water and lots of sun. Tennis courts and fishing rods are most popular forms of entertainment.

Our luggage finally arrived, and we are in clean clothes again. The trip down, while beautiful, was unbelievably dirty. We saw some wonderful scenery in southern Oklahoma and Arkansas, but Louisiana was most unimpressive, especially Shreveport, and much of Texas looks just like Kansas, although I prefer the latter. The sea breezes, though, are pretty swell and make this Texas heat bearable.

Sunday we move to an outlying field to begin 3 or 4 weeks of basic in SNV's. We get about 35 hours of various kinds of flight and then return for instrument squadron.

June 27, 1944

Dear Phyl,

Day number 2 at Corpus—strictly Snafu'd. Our luggage hasn't arrived yet, and if it doesn't come by 1500, the chances are that we'll be held over an extra week, not a pleasant prospect. Here's why. Someone at Olathe wasn't on the ball, and most of us packed our log-books in our bags. These arrived at the Union Station late and missed the train. So we've been living in our one set of dirty khakis for two days now, and then they pull this on us.

NATC Corpus Christi is a whale of a big base, some 20,000 acres and lots of beach front. There are 3,000 officers, 7,000 cadets, and nobody knows how many enlisted men and women here. Barracks are fair, six men to a four-man room, but accommodations are for 2,000 instead of 7,000! There are all sorts of recreational facilities if you can find them. My first act during free time yesterday was to find the barber shop and get a crew cut. It looks like the dickens, I'll admit, but it is certainly the right hair style for this hot Texas climate.

Lots of planes flying around, mostly PBY's and OS2U's, as training squadrons in both are located here at the main base. The big P-boats look nice taking off out on the bay. The program here starts out with 3 or 4 weeks of Basic in Vultees at an outlying field, with classes in Navigation, Engines, Code, etc. Then a week of All-Day Ground School, at which time we choose a branch for specialization. Then instrument squadron followed by specialized.

The washout rate is interesting. 17,000 are to be eliminated. 30% of those in or below Preflight will wash and 50% of those in Primary. Here, a minimum of 15% will go. The last group at Olathe were asked to quit. Very encouraging. However we're too used to that sort of stuff to get very excited and work like beavers, but somehow it just doesn't seem to make much difference. That may change when we start flying, though.

Mess here is interesting. We are seated eight to a table and served by colored mess attendants. All very formal, but everyone has to quit at the same time, and so you eat rapidly and not very much. The chow is pretty good, but we would all prefer the enlisted men's mess.

There are lots of eager officers around here, and even the cadet officers are eager to show their rank, but we'll change that in time. Demerits are easy to get, and 50 will wash a man out. . . .

Flash: Luggage just arrived—we leave Sunday for basic! At last I can start living like a white man again.

June 29, 1944

Dear Phyl,

Another day about over—and what a day. This morning we had a three hour lecture on aviation safety devices, and this afternoon and evening we are secured, so I put in three sets of tennis and an hour of swimming. This is the best place in the world to pick up a sun tan; you can't go out half an hour without either burning or tanning. I'm doing a pretty good job of the latter.

Yesterday we spent a hot and sleepy morning listening to a lecture on chemical warfare. The lecturing officer was awfully eager for gas warfare and gave us all the gory details. Then in the afternoon we got another one of the umpteen physicals—this time a very thorough one. I got past it O.K., so I guess I'm still a pretty fair specimen after three weeks' leave.

Tomorrow we get the low pressure chamber—simulated high altitude flight. We go up to 18,000' without oxygen and then on to 30,000' with masks. I'll let you know how it goes.

June 30, 1944

Dear Folks,

This week of indoctrination is going by very nicely. We've checked out a lot of gear and books and had a number of lectures and movies on air safety, chemical warfare, etc. Yesterday afternoon we were secured, and so I got in about three hours of swimming and tennis. In this Texas sun that means the addition of several layers of tan. The sun down here is even more potent than it is back in good old Kansas.

We had another physical exam, same as when we enlisted. Very thorough, and I was quite happy to get through it successfully. This morning we went through the low pressure chamber. This was quite interesting. The purpose is solely to show you the effect of high altitudes with and without oxygen. Fourteen men are placed in the chamber, about 30 feet long, eight feet high, and twelve feet wide, with two corpsmen attendants. Pressure is reduced to 5,000' and then dropped to 2,000' to test each man's ability to clear his ears, then a steady climb is made to 18,000' without use of oxygen. On the way up you don't notice anything other than constant popping of the ears. However after five minutes at 18,000', you notice light-headedness and signs of dizziness. By the end of fifteen minutes I was quite glad to get on my mask, and the effect of oxygen is wonderful. Vision improves about 50% and you feel swell physically. From there up to 30,000' there's no appreciable change. At 30,000' one of the fellows removed his mask and tried to write a letter. He lasted 3 minutes and then was out on his feet. He lost conscious control after about two minutes. The descent wasn't bad, but you have to keep working on your ears to keep them from stopping up due to increased pressure. All altitude changes were made at 3,000' per minute, about the rate of climb of a fighter. Tests were given at sea level, 18,000', and 30,000'. With 225 as perfect, I got a 210 at sea level, 107 at 18,000' without oxygen, and at 30,000' with oxygen, a 225. It really increases your vision and mental activity. All night flying now is done with oxygen, regardless of altitude, and vision will increase 40–60%. The test was marking out one letter in any given three-letter combination. At 18,000' I found it quite difficult to concentrate and had a little difficulty at first in remembering just what I was supposed to do. With oxygen it was a snap, and the last test was the only one I was able to complete.

It is an unceasing pleasure here to run into fellows you haven't seen for some time—from Liberty, W.T.S., Preflight, and those who left Olathe last week. The boys in the draft in which I was supernumerary are now at Cabaniss Field and are all soloing. We are not sure yet where we'll go Sunday, but all indications point, rightly, I hope, to Cabaniss.

There's lot of talk going around about whether we'll strike for Navy or Marine Corps and what branch to get into—carrier-based, land-based bombers, patrol boats, or VO-VC cruiser-based observation. I'm all for staying with the Navy, but am decidedly undecided as to fighters, land-based bombers, or P-boats. 70% of each class goes to fighters— CV, about 15% to P-boats, 8% to VB2, and about 4 men from each class to VO-VC. However I'd better get through Basic before I start losing sleep over it. . . .

June 30, 1944

Dear Phyl,

Well, we're all checked out now and have the afternoon and evening off. This morning we heard several lectures and then attended graduation exercise. You hear a lot about graduation—formations flying overhead, hundreds of cadets, fond papas and mamas, and beautiful girls. Sounds great, but isn't quite accurate. It's as hot as Texas; the audience consists of two hundred sweaty and disgusted new cadets and about twenty-five parents sitting in the sun. After ten minutes waiting, the Admiral arrives and delivers a short, pointless address upon the importance of naval aviation, stressing the point that the graduates are Navy aviators and not Japs or Army pilots. After the designations are passed out, the new officers give a cheer and the ceremony is over. The observing cadets march off, wiping perspiration from their brows, muttering under their breath, "Those lucky b——s." That's graduation at Corpus. When the Admiral—Mason—spoke to us, the high point of his address was the question, "Why am I Commandant instead of you?" The obvious answer was given, "Because I'm older than you are in length of service." With this deep thought into which to sink our intellectual molars, we spent the afternoon playing tennis and swimming. It amazes me that I can still burn; I'm no longer tan, but rapidly becoming light brown.

Highly incidentally—on the trip down I learned and perfected the sailor's method of lighting a match in high wind. It's foolproof. . . .

Later. Had a wonderful steak dinner at the Recreation mess tonight. Funny what the sun does to hair—mine, I mean—I look like a silver fox, black beneath and silver tips. My buddy who is a blond Californian is trying to get me to peroxide my hair and be a blond. Sounds good, for a change. . . .

Tomorrow we move out to Cuddihy Field and will probably fly on Monday. Cuddihy is known as "Country Club" and is supposed to be easy on military and ground school but tough on flying. We'll see about that. Had some entrance exams last night, and it seems I actually learned a little about Engines and Theory of Flight. Ground school in Basic includes only Engines, Navigation, Code and Blinker, and Gunnery.

Cuddihy Field
July 3, 1944

Dear Phyl,

What a day! What a wonderful day because I find four letters from you in the mail box. And what a lousy day because I got my first ride in the V. So help me, by the time I screwed up half a dozen landings, forgot all about flaps and prop pitch, I was ready to turn in my goggles. No kidding, right now I feel worse about this flying than I did in Primary A with one down check against me.

July 4, 1944

That was no way to start a letter, so I'll try again. The above was true yesterday, but I've regained a bit of determination and confi-dence. . . .

As I mentioned, I had a little trouble yesterday on my first flight. This wouldn't worry me but for the fact that I only have two more duals before the first check. Older cadets tell me, however, that second and third duals go much better than the first, so if I get on the ball and stay on it, I should get the job done. Flying here at Basic takes only two weeks. Our ground school at present consists of navigation and engines. The third week, all we have is gunnery. P.T., of course, all the time.

Cuddihy is a beautiful little base, the most compact I've ever seen. The ground school and mess hall are both about fifty yards from my room, and the gym and hangars about a quarter of a mile, which is very nice after hiking all over the main station for a week.

July 5, 1944

Dear Folks,

Methinks it's high time I got off a good letter home. Should have yesterday, but I wasn't in a very gay mood, so refrained from sending home a poor report card.

I guess I'd better start with Saturday of last week. In the morning we heard about four hours of repetitious lectures and witnessed a graduation ceremony. It was quite a let-down. I'd always pictured thou-sands of cadets, parents, gold braid, twenty-four plane formations over-head, and a swell band playing "Anchors Aweigh." Actually there were 225 boot cadets (us), and about 25 spectators; one lone SNJ flew over. The Admiral (Mason) made a very poor and lengthy speech; the newly commissioned ensigns and 2nd Lieuts. gave a feeble cheer; the band was sad, and we were soaked with sweat (you don't perspire here, you just sweat).

We were free all afternoon and evening, so played a hard game of tennis, added a couple of coats of tan, swam, and then cleaned up for dinner. Three of us ate at the Rec. Mess—real dinner steaks, the like of which for size cannot be matched in civilian life, but not nearly as delicious as the toughest steak Shreve ever sold you and mother cooked.

Sunday morning we went to church, packed, sent out laundry, and moved out to Cuddihy Field to commence Basic. And I mean we began at once. Arriving about 1200, we were in lectures from 1300 to 1915, took code tests, cockpit check-outs, got books, and were so tired our tails were really dragging.

Monday was no better; Navigation, aerology exams, power plant check-out, and first dual flight. Here, when you're not flying, they keep you as busy as possible in more classes and lectures, instead of letting you relax as we used to do.

I drew a fine young instructor, but my first ride was sad. To say the least, I doped off and merely went along for the ride. However, it put me on the ball, familiarized me with the plane and area, and today's ride was a decided improvement. But I'm ahead of myself.

Liberty here at Cuddihy comes at Monday night secure and all day Tuesday. Being a bit flat of purse and not feeling I'd like an unplanned liberty, I stayed on the base, went to a show, and slept until 0730. After 0800 breakfast, I studied course rules, navigation, and did some reading the rest of the day, resting up and recovering my badly shaken self-confidence after the first ride.

As events today proved, my time was well spent. I had no trouble with the Navigation exam, had a good flight, and feel fine. Contrast this with the lot of one of my roommates, an old friend who was in the 5th Batt at NFPS and went to WTS, NFPS, and Olathe with me, who went to town, had a swell time, flunked the nav. test, and at present is stuck in the mud in his plane at an outlying field.

A word about the SNV. Built by Vultee Aircraft Corp., it is a low-wing, all-metal, service type craft, powered by a Pratt & Whitney Wasp Jr. 450 HP, nine cylinder radial engine, with Hamilton controllable prop—two pitches, high and low, and equipped with a two way radio set. Considerably more airplane than the N2S. There are about twice as many instruments to watch, plus a different type of elevator tab, rudder tab, and flaps. Keeps one kind of busy. The V cruises at 120, climbs at 90 at 400–500 ft. per minute in low pitch with 20° flaps. Top speed in level flight is about 170, and it is restricted to 230 in dives. We do no acrobatics or violent maneuvers in them outside of demonstrated stalls and spins which may not even be practiced solo. Cuddihy has seven practice fields, all lying west of the base, which is itself almost due west of Corpus Christi.

Cuddihy is the nicest, if the toughest, Basic field in USNATC. All drives are asphalt, and the grass is really amazingly green for Texas. To show the compactness of the arrangement, the ground school and mess hall are both less than 100 yards from my desk. The barracks are quite nice, four men to a room.

SNV Vultee Valiant flown in Basic training at Corpus Christi (Courtesy U.S. Naval Institute, Library and Photographic Services)

Reveille sounds at 0545. Breakfast at 0600. Morning classes include an hour and 40 minutes each of Nav. and engines, and an hour of P.T. which is a farce except for 10 minutes of violent calisthenics. Noon chow is at 1225, and we are at the squadron hangar from 1300 to 1830. Tonight we left the mess hall at 1950, so you see, I have a pretty full day. Gotta study now. Write soon. I have my first check ride tomorrow.

July 7, 1944

Dear Folks,

News at last, and happy am I to say that it's good. As we ran into a rainstorm yesterday afternoon, I didn't check until this afternoon. It wasn't too red hot a ride, to my notion, but the check pilot was satisfied and gave me an up. This means that I have no more check rides to worry about for all of two days. I get four solos and then get my A-9 check, which is exactly the same as the A-4, only much better, I hope. Should check on Monday if all goes well.

The past two days we've been swimming in P.T., passing our AAA tests—mile swim and quarter-mile rescue tow. Neither was as bad as I'd anticipated, but I was plenty glad to get back on my feet. Swimming gives me a stiff neck. That's all the swimming except for the maintenance test, which is swimming the 440 in ten minutes. That's a little rough but can be done.

In navigation we have begun to use the sextant. It's quite an art to use it rapidly and accurately, but the mastering of that art is a key to long life as a navigator. There's really nothing complicated about it, but it takes a very steady hand. I can imagine what it would be like in a plane on a rough day. We've also begun use of the Army D.R. [Dead Reckoning] computer, a small wind, course, relative movement plotter which supplements the standard Navy plotting board which I've shown you. It's a handy little gadget and ideal for short hops or contact flying, but not too suitable for long distance work where positions are important. . . .

July 9, 1944

Dear Folks,

Just a note before chow to let you know that all is going well with me. This morning I flew an up on my A-9 check, which is supposed to guarantee that you'll get through Basic. I had a pretty decent check pilot and flew him a good ride. Now I have ten formation hops and three night flights—a week of flying at most.

Ground school is coming along in good shape. I got a 3.8 average on my Nav. entrance exams, which was way above the class average. The first test in Engines wasn't bad, so everything here at the base is going along smoothly. . . .

Among the new boots here at Cuddihy is one of my Iowa City roommates—Ed K—— from L.A.

[July 11] (This is Tuesday) So this morning he and I, refreshed by a long night's sleep, went into Corpus to see the sights. I wasn't greatly impressed, although it is a nice little city. We took a boat trip out on the bay, had dinner at the Driscoll Hotel (city's best), which was quite a pleasure. Good food and a lovely dining room. In the afternoon we went over to the main base and saw a lot of the old fellows from Primary, Preflight, and W.T.S. Lots of fun talking over leaves, washouts, and the latest scuttlebutt.

Tomorrow I start flying formation, which should be a lot of fun. Our formation area is pretty big, and it will be a lot more fun than shooting landings for an hour and a half. . . .

July 13, 1944

Dear Phyl,

What I'd give for some time to study and write letters. Ain't none, though, so I'll start this and finish it sometime. First off, all goes well with me. I got an up on my A-9 check, which was quite a relief, and have had two of the ten formation hops. Tonight I get in my night flying and should finish flight Sunday. Ground school goes along sleepily. We have the most lousy nav. instructor I ever saw, but my average is still a respectable 3.8. Engines is kind of rough, but I'm not worried about it yet. . . .

July 17, 1944

Dear Folks,

. . . Yesterday we finished our classes in Navigation and Engines, although there is a Nav. test next Sunday afternoon. Our ground school class now is gunnery, always an interesting subject, especially when your life may depend on it some day. At last this cadet existence is coming up to par. For the past thirteen months we've heard about the swell life of cadets and seen very little of it. One of the items on the "ideal" cadet's curriculum was gunnery in the form of target practice. So we froze our hands and dirtied our clothing lying on the range at Iowa, had our nerves and eardrums jarred considerably by .45's at Olathe, but now we catch the bus at 0845 every morning, ride out to the range which is landscaped like a golf course and as clean as the brightwork on a battleship, and with the best shooting equipment available blaze away 50 rounds of 12 ga. shells a day for four days. That's more, I'm sure, than Dad and I have used in the past 12 years. Our shooting includes both trap and skeet, the big thing being to teach us the importance of leading a moving object. We use Remington 12 ga. "Sportsman" automatic shotguns that hold 3 shells, and shoot #8 shot with special tracer shells in case you can't see why you're missing. Today we shot our first 50 rounds from two of the skeet positions, both high and low birds. The 12 ga. is just a little bigger than I like, and the amount of lead at short range is amazing, but I did very well, so I was told by one of the instructors, breaking 39 of my 50 birds. With any kind of practice one should be able to do very well at trap shooting. I'm anxious to try my hand at regular skeet. However, my right shoulder tells me that there's some difference between shooting 5 rounds with a heavy jacket on and 50 rounds with only a skivvy shirt to absorb the kick.

So much for ground work. Today I completed my flying and have put away my helmet and gloves for possibly a month. There will be this week, possibly next week as a stand-by, a week of all-day ground school, then possibly one or two weeks wait in instrument pool, plus some time in a Link Trainer before I'll fly again. But after the last five days of really grueling formation flying I'm ready for a rest.

In addition to continuing the form[ation] work done in Primary, we learned the step-down positions—used almost exclusively in the fleet, I understand. To [add to] our troubles in flying good positions and making crossovers, the weather has been quite turbulent, and yesterday we flew in and around thunderstorms. That was quite a day. After 3 hours your back and neck get stiff and your feet almost go to sleep, and yesterday we flew 4.5 hours. I mean we were dead tired when we got down. Our second ride was most enjoyable as we climbed to over 8,000 feet dodging in and around rain clouds that went up to 20,000', and flew over Alice, Tex., airport and Kingsville, big Navy fighter training base about 40–50 miles north of Cuddihy. Combined with our form[ation] hops is a little unofficial tail-chasing which is as close to aerobatics as we'll get for a long time. It's a real thrill to stick the nose of a V down about 60° and let the airspeed build up to 160–170 and then pull her up in a climbing turn and find yourself just above and behind the lead man.

There are now a hundred or more SBD Dauntless dive bombers here being tuned up for training use in the near future. Instructors are being checked out in them every day. Our squadron, 12A, is being disbanded, and all cadets will now go to Cabaniss for Basic. Cuddihy is to be turned into a Pre-Operational base by September. That means another month or two or three as a cadet. The instructors are plenty sore too, for the Navy is kicking out a lot of the old school teachers and insurance salesmen who were given commissions for training purposes and to man newly commissioned vessels, and changing A(V)N officers (flyers) to A(V)S (specialist) and giving them ground jobs and routine deck duty on ships unattached to any branch of aviation. Paul S—— wrote that two of the fellows from Friends got the ax at St. Mary's [Preflight] in the 50% washout deal. We are surely the boys Uncle doesn't love unless he needs us.

Even P.T. is nice these days—except for those calisthenics—played tennis today for an hour during regular class time. Still we are unpaid, and I am flat broke—yea, in debt 25¢, having put some ten bucks to usury about three weeks ago when I was flush. Scuttlebutt is we'll be paid tomorrow. I hope!

July 22, 1944

Dear Folks,

Write this off as another base successfully encountered and mastered. Today was our last regular day, with finals in Navigation and checking out at Squadron. The navigation finals were pretty rough. I 4.0'd the celestial, but got only a 3.3 on the D.R. which had some bad stuff in it where the width of a pencil dot would throw you off as much as four degrees. Answers in a class of 60 on one bearing ran from 305° to 025°, which is quite a divergence of opinion among fellows who've been doing that kind of navigation for over a year. In view of my good marks in celestial (4.0 on both entrance and final) I'm wondering whether I should put it to use in P-boats or VB2 instead of CV where it is never used.

Gunnery ended up well with a thorough examination on military weapons, or rather small arms, that netted me a 3.6 for the course—your shooting doesn't count. Shot skeet twice more with interesting results of 33 out of 50 both times. I find I can hit about 23 out of 28 singles shots, that is one bird at a time, but the doubles, two birds at once, drop my score from good to average. We had an exhibition of real skeet shooting put on by one of the Chiefs and a 1st class gunner. The Chief used a Winchester repeater and the 1/C a Remington over-and-under that was a sweet looking gun. It was amazing to watch them blast away and get hits every time. Their birds never traveled over 20 feet before they got off their shots. The Chief hit 25 birds with 26 shots, which is good enough for anyone. Of course he's been shooting every day for over two years, and practice doesn't hurt. The way the shells fly on the range still opens my eyes. Our one class will shoot up 2000 rounds in an hour, and they used to have four classes a day, fully twice as large as ours! Now you can see why the rabbits in Kansas are a little safer this year. . . .

July 29, 1944

Dear Folks,

Another week is over, and tomorrow we go in to the main base for the week of All-Day Ground School and probably a week or two of instrument pool. It seems like I haven't flown for months instead of a couple of weeks, and it will be at least two, and maybe three more weeks before I get up again. . . .

I especially appreciated Dad's letter and enclosure as the week has been a little strenuous on the old billfold, as I've been on the base a minimum amount of time. Our scheduled activities have been P.T. and code for those who didn't pass the tests (which doesn't include me). The rest of the time we were on liberty and in Corpus. Consequently I got pretty familiar with the town and environs—such as they are. The week was greatly enhanced by the fact that Phyllis came down, and we had a swell time in what might otherwise have been a dull week.

I find the most interesting part of Corpus to be the small-boat basin with its pretty good assortment of motorboats, launches, and sailing craft. Spend quite a lot of time down there boating and sailing and watching the gulls. These are practically tame and will gather almost by the hundreds to eat peanuts tossed to them by people along the sea wall. When I reach instrument pool, I intend to learn a bit about sailing myself, as the basin is ideal for it, and I haven't seen anything around Corpus more interesting—certainly has it all over a motorboat of any size or variety.

One afternoon I wanted to do some driving, and in the process of getting a car, was faced with the fact that my Kansas license was over a year expired. At the Texas License Bureau I learned that all red tape isn't in the Navy. First there were forms to fill out at great length, then a booklet to study, supposedly at great length. This preceded a test which was more formidable in the minds of the Texas Rangers, or whatever they were, than it was on paper. Then came an eye test and a test on road signs and rules. Passing all these, I had to secure a car and driver and take a driving test that lasted half an hour. It was quite an ordeal, but I had to sympathize with one poor sailor who only had an A [rationing] card and had to use his gas to get a license. At least I now have a license good for two years. Car rental down here is very reasonable, and you can drive all over Corpus without picking up any more miles than if you drove across Wichita.

On Friday we were informed that we had to fill out our Navy Preference Sheets at once, instead of at Main Base, so I sweat out a lot of arguments with myself for a couple of hours Friday a.m. The result was that I signed up for VB2, land-based bombers (same thing Bill Roy's in) with CV (carrier duty) second. If I'm very lucky, I'll get the VB2, but as 70% of each class gets CV, I feel I'll get it anyway. Either will be alright. VB2, should I get it, will mean a long, tough grind, the hardest down here, flying the SNB's, Beech, made in Wichita. In operations it might mean anything from P-38 photographic reconnaissance to the new single tail Navy B-24, or a fast medium bomber, my choice, or perhaps transport duty with a VR squadron like that based at Olathe; a final chance would be getting duty as an airline pilot with both civilian and flight pay—unbelievable, but it does happen. Well, that's what I put in for—big stuff and multi-engine, now to see how long it takes them to make me a buzz-boy in a Hellcat. That wouldn't be bad either. I should know which it will be in a week or two.

Had a swell letter from Marvin S——— at G. Lakes. The lucky stiff is already a Seaman 1st, company clerk, and it looks like he'll do alright, although he gripes about discipline. Who doesn't. I've seen several of the old gang from Olathe in town and have been getting the word on ground school. The big job is supposed to be staying awake.

If I spend another six months down here I'm going to learn this Mexican lingo. It gets my goat to listen to 50% of the population talking at a great rate and not have the vaguest idea what it's all about. At least the little bootblacks know enough English to say, "Shine Cadet?"

Main Base
July 30, 1944

Dear Phyl,
New quarters and lots of old faces here at the main base. We came in this afternoon and checked out books, then secured for the evening. I've seen a number of the old fellows from Primary, Preflight, and even W.T.S. Tomorrow we get the pressure chamber hop and an afternoon of lectures on aviation safety or something. Regular classes will begin Tuesday.

We have a nice room, but that's about all that's good. There are six of us in a 4-man room, and our barracks is the last one in the area, which means that we are about a half mile from everything. Quite a change from Cuddihy.

August 1, 1944

Dear Phyl,

. . . After two days of this ground school I'm ready for another week of liberty. . . . We have a pretty full day—four hours of Radar, two hours each of Engines and Aerology, and an hour of Recognition and P.T.

To make things better, we are nearly a mile from the ground school and have it to walk six times a day. The big job is staying awake in class. Nothing seems too rough, and some of it, Radar particularly, is very interesting. Wish I could tell you something about it, but it is quite confidential. We can't even take notes in order to study. The place is guarded constantly, and we have to show our I.D. cards to go to class.

August 5, 1944

Dear Folks,

Only one more day of ground school, and I'll be very happy when it is over. These are the longest days I ever put in. Mornings aren't bad as we have an hour of P.T. and 40 minutes rest period to break the monotony, but the afternoons from 1300 to 1900 are just too long.

P.T. isn't bad—so-called aviation safety—which in this case means familiarization with life rafts, life jackets ("Mae Wests"), abandon ship procedure, and regular swimming. I really like the life jackets. I'm light enough that I can just relax and float; could even sleep if need be. Being small has its advantages too with a one-man raft. The big boys have quite a time getting six feet of cadet into a tippy four-foot boat.

I'm learning a lot of amazing things about Radar. It's really wonderful stuff—wish I could tell you about it, maybe some day it will no longer be a military secret. Tomorrow we finish up and have tests in Radar and Aerology.

The list for specialized training finally came out this evening. Practically the whole class got CV, including myself. I am not particularly sad about it, as is one of my buddies who wanted P-boats and got CV—he's seriously planning on quitting. I'll be here for another week and perhaps two. We'll probably go to Chase Field at Beeville for instruments, but I may join the glee club and get myself into the Main Base instrument squadron. Time will tell. . . .

August 6, 1944

Dear Phyl,

Your recent letters have done much to raise my morale during the last few days of ground school, which was a good thing, as it was darned rough. Hot? Second deck, all the windows closed, and the fans didn't work. Six straight hours of that, and you don't have a dry stitch on you. Anyway, it's all over now. The Radar test was a snap, the Aerology was the roughest I've ever seen. Practically everyone used a gouge on the Aerology, but I honored my questionable integrity and intelligence and took the test. Wow! I mean I really sweated it out. Still don't know how I came out, but I am not losing any sleep over it. . . .

We are beginning our first week in instrument pool. Scheduled activities consist of an hour of communications and an hour of P.T. The rest of our time is free, but we have only one overnight, and daily liberty doesn't begin until after all activities are over, which may be morning, afternoon, or evening. At least we will get a lot of rest, sun, and letter-writing. It's funny how a tan will disappear down here as fast as it will come.

I'm planning on getting a ride in a PBY tomorrow to see just what the big boats are like. Everyone gets a ride while they are here, and I feel like skipping P.T. tomorrow already today. . . .

We're "getting the word" on CV advanced. Here at Corpus you have to qualify in dive bombing, strafing, and fighter tactics. That's a bit rough when at Pensacola the boys specialize in just one branch—fighter, dive, or torpedo bomber.

August 8, 1944

Dear Folks,

Not much to write about now that ground school is over, but I am thankful for the chance to rest after last week's ordeal.

Yesterday in P.T. we took our final swimming test, the maintenance test of a quarter-mile in ten minutes. We also had more tests in code, blinker, and semaphore at top speed, ten words per minute, but the code baffled me. The test was sent in regular message form, heading, address, priority, time and date, etc., and then the text in odd groups of 5, 4, 3, 2, or 1 character groups. The final third of the test came at 16 w.p.m. in plain language and was just too much for me, as I haven't heard any since W.T.S. days when I was able to take 16 w.p.m. in plain language. However, we don't have to check out in it here, and with a little practice I know I will be able to get it without any trouble.

This morning I took a two hour hop in a PBY, just to see what it was like. It was a revelation, and I am very glad I didn't get assigned to them for Advanced. They vibrate like an old Ford, are as slow as an N-2-S, and when they land, it's like a house hitting the water, there's a terrific splash and jar, and then a series of jarring bounces and groans from the members of the fuselage. A big thrill in a PBY is when you make a 30° bank or dive at 100K. That's not for me, and I'm happy with the prospects of a Hellcat or F-7-F. . . .

Chase Field, Beeville, Texas
August 13, 1944

Dear Folks,

Well, here's "the word" again. I'm out of the pool and into instrument squadron 13B at Chase Field, Beeville, Texas. Which, geographically speaking, means I'm some 60–70 miles north of C.C. on what might definitely be termed Texas plains. Two things are immediately noticed: first it is hotter, and second drier, much like Kansas in the summer.

Having been here but a few hours I can't say much about, or for, this base. It is the worst I've seen as far as physical equipment goes. The barracks are flat-sided and topped, covered with tar-paper of a dirty dark green hue, and stripped to the barest essentials. In place of the beautiful grassy lawns of Cuddihy, we have a few spots of sick looking crabgrass and lots of dust. It's a sad looking place. Corpus is over 60 miles away, San Antonio over 100. Beeville is decidedly not a liberty town. If all goes well, I'll be here five weeks, or perhaps only four. We'll get work in Link Trainers for the first week or so and then fly V's under the hood, that is, blind flying. Some of us will get SNJ's, but no one knows how many or who as yet.

The last few days in the pool went by smoothly enough. Played some tennis, almost got another P-boat ride, did some reading, had some dental work done, and still a couple of fillings to go. If these Navy dentists can't find a new cavity, they take out an old filling and put in a new one. The dentist at the main base was swell, but I don't like the looks of the one out here. . . .

Speaking of Texas, the ants here at Chase are so big and so numerous they have their own roads and sidewalks! The address is still the same, do write soon.

August 13, 1944

Dear Phyl,

. . . As I stare out my window, my eyes are greeted by the sad sight of Chase Field, Beeville, Tex., where I am hoping to learn instrument flight during the next four or five weeks. I say it's a sad sight and it most assuredly is. The grass is burnt to a dusty brown, the barracks look like overgrown boxes and are covered with tar-paper. The interior of the mess hall strikingly resembles a barn, and I've seen barns in Kentucky that would make better living quarters than those we call "home." We are 60 miles inland, and the once-cool sea breeze is now coming directly off the hot plains—deserts would be more exact—covered with cactus and mesquite. The spirit around the base seems to reflect the physical characteristics. There is an utter lack of enthusiasm, and though the course is tough, no one seems to care enough about the outcome to assume either an active interest or worry. Corpus is 60 miles south, San Antonio over 100 miles north, and Beeville is just another Texas cattle town, half-filled with Mexicans.

That is a fine, condensed version of my current bitches, so I am embarking on a strict program of self-discipline for the stay here, to take my mind off the surroundings. This includes intelligent flying, rapid and accurate navigation, a limit of two packs of cigarettes a week, no beer (unless you want a letter like the last one), and as much reading as I can get done. . . .

I had a delightful time Saturday in Corpus. Met a cadet from South Carolina and we decided to go sailing. So we rented a little snipe boat and with yours at the helm, cast off with questionable expectations. All went amazingly well and we had a great time, although it did take us half an hour to get the boat alongside the dock. The only casualties were the seats of our trousers which absorbed salt water and blue paint in some quantity. We sneaked up to the Driscoll for dinner to the tune of the shine-boys' comments, "Hey, cadet, the seat of your pants are wet!" Great.

August 16, 1944

Dear Folks,

　. . . This being our day of liberty, I've been taking it easy. Would have gone into Beeville but am missing my hat and so can't get off the base. So I have located all the base's facilities for recreation and used them, including the library.

　After two days of work, I'm beginning to get onto things and think I'll find it a fairly interesting if not especially enjoyable base. I got off to a 4.0 start in navigation, and we finally took up something new in D.R. nav., so my interest is aroused again. In class we daily get a short test for speed practice, and a long test for grade, and another long problem for homework, due the next day. So far I've gotten the homework done in class, which lightens the load and my mind considerably. Other ground school subjects—Engines, Recognition—are simply dull. Same stuff we had in NFPS, and we know it no better, if as well. One thing has struck me here—the g.s. instructors are the wildest bunch I ever ran into—all characters of great individualism.

　Yesterday I had my first Link trainer hop and enjoyed myself immensely. I believe I did creditably well, though far from perfect. The Link isn't a plane—merely has the same controls—very mechanical and stiff. Under the hood on instruments you do get some funny sensations and certainly cannot rely on any sense of turning, but thinking and watching the right instruments you can do pretty much whatever you want. It's amazing how much I've come to rely on senses, both hearing and touch as well as sight, in flying, and eliminating these, you have only your mental "sense" to go by. It is said that 40% of this is lost the moment the hood goes down. This I don't believe is going to be my case and I intend to make a good Link flier. The problems of radio range orientation are very interesting and themselves quite simple. The trick being to work the problem and fly the Link at the same time. Incidentally the Link Dept. is run by Waves and they do a very good job of it.

　But heaven help me if I try flying a plane like I do a Link. I think it will be a lot easier—I hope. I was fortunate to be assigned a flight board flying SNJ's (North American's AT-6 "Texan"), so will not be new to them when we really start flying again in Advanced. The only drawback is that my flight group takes a week longer than the others and is a little tougher. They're giving lots of downs out here, but a very low washout rate—they say.

　There's not much that can be said for the base. It remains ugly and uncomfortable, but not as bad as it might be. Reminds one very much of Flight Prep or Preflight—but I lived thru them, why not this? . . .

August 18, 1944

Dear Phyl,

. . . Events at Beeville move slowly but steadily on. I had my last practice hop in a Link today, and tomorrow we begin work in earnest on radio range procedure. It will be very interesting—I just hope I can get on the ball and stay on.

Which reminds me—you asked about the meaning of "gouge." It is a naval term meaning to cheat, crib, copy, etc. "The gouge" is a copy of the answers to a test. Of course it is highly frowned on, and if caught, a gouger is usually laundered. So I see little reason for taking the chance when I can do well enough on my own.

I am, frankly, having a devil of a time with my navigation. Of course my roommates don't see my point when I have a 3.7 average, but I seem to have lost the knack for D.R. plotting board work, and it's really got me worried. I was much better at it ten months ago than I am now—which isn't good any way you look at it.

We got another circular letter today stating that the program is tightening up again—as if we needed to be told—and that pre-operational of 3 months will definitely be established, so God only knows when I'll get out of this. Downs are thick around here, so I'd better worry most about instruments and not pre-op.

August 21, 1944

Dear Folks,

Not much news, but I'll see what an ex-editor can dig up for copy. We have seen a rare and wonderful thing in Texas these past two days—rain. Not much, true, not enough to dampen the ground, but it reached the ground at any rate. All this is a result of a big low-pressure area moving north off the gulf, bringing every kind of cloud formation, wind shifts, and poor flying weather. Aerology reports a hurricane 36 hours away; the station is on Condition 2; special hurricane watches are posted; all liberties are cancelled, and every cadet is positive there is no danger of the storm coming our way—we've seen too many false alarms to get excited now. At least it gives us something to talk about. Should the storm come much nearer, all planes will be flown to an Army field near Dallas, piloted by all instructors and some forty cadets who have finished flying. In regard to the planes, this is probably a good precaution as there are no facilities here for the slightest protection. I still don't see how they operate this squadron with just one hangar for repairs and overhaul.

I still haven't flown and don't know when I will. The Link work is coming well enough and is increasingly interesting. I just wish I had twice as much. Fifteen hours isn't very much.

Yesterday a new group came in, and we've had a lot of fun giving the boys the word. Five of my good friends were in the new group, so it is more like the old times. Also Sunday we had our final final on ship recognition. I pulled out a 3.8, so feel pretty good about it. Tonight our ballot requests were distributed and will be gotten off soon so the government can get the necessary papers to us prior to 7 November. I still want a letter from Dad on his views as to the respective merits of the candidates. I heard a little of the Democratic convention while at Cuddihy and was not happy with what I heard.

I am beginning to catch on to the morale situation around here, and it is far from good. The story is something like this. All of us have been in training for 15–20 months. In that time the program has nearly doubled in length and the number of washouts increased immeasurably. We have taken a lot of unnecessary and highly unpleasant riding from officers whose length of service is in many cases no greater or less than ours and who are not mentally or physically qualified to step into our shoes. We have seen a lot of our good friends washed out for the most minor reasons—men who would make good pilots and good officers. We also see a thousand Lieutenants in the A.A.F. who don't have as many hours as we do, but are flying P-47, P-51, P-38's. Or just say they are flying. By actual count, I've been here eight weeks and have flown two weeks. Thirty hours in three months is hardly a drop in the bucket. Our training schedule is such that we are terrifically overloaded for one or two weeks and then have nothing for a longer length of time. We are potential officers, but so far from it that we are lower than an ordinary seaman when it comes to general social recognition and treatment.

Now we are told that the Navy no longer wants pilots and is trying to wash out as many as possible. This is now hitting us here at Corpus, though still to a lesser degree than Primary. 30% of the washouts at NATC are from this squadron, and the entire cadet regiment celebrates when someone gets an up-check. The three months pre-operational is now official, so the end is yet far from sight. And because of the present surplus, boot ensigns are being made ground instructors, personnel officers, or washed out completely and sent back to be drafted into the army. I doubt if there is a man in the squadron who will say he expects to get through—it's too much on luck, and she is a fickle mistress. No one blames the ex-cadets at Great Lakes who booed an Admiral off the platform when he tried to address them on their place in the Navy. These same cadets have effectively refused to salute Chief Petty Officers there, and threw one out a window when he attempted to assert himself. These are all fellows like myself—the only difference perhaps that I got a Santa Claus and they didn't. No one looks down on a washout, just thanks his lucky stars he hasn't gone the same way yet. If there were the slightest demand for our services, for the $30,000 the Navy has put into our training, we'd work like mad, but we are constantly reminded we are almost undesirable and would be much more valuable as un-trained boot seamen. All this for the dubious honor of wearing gold wings. It's no wonder that the most talked of scuttlebutt is that about September 1, the Navy may release cadets to civilian life if they wish. The above doesn't include any of the personal unpleasantness in many phases of the program, and facts stated are not scuttlebutt. Anyway, that's how we look at it from the inside. The old salts say that if you can't gripe it isn't a good Navy, so there must be something good someplace. Time for taps, got a lot of work to do tomorrow.

August 22, 1944

Dear Phyl,
    . . . We are having a little change in the weather. It seems a hurricane is moving in from the Gulf and with it is coming every type of cloud imaginable, and best of all, rain, real rain. The hurricane has been scheduled less than 24 hours ahead of Beeville, so the base is in Con-dition #2, or possibly by now, #1, which means that everything is secured at the flight line, and if #1, the entire base will be secured, no scheduled activities held, all liberties and leaves cancelled, everyone stands by to turn out in case of an emergency. The planes will be flown

to Dallas by instructors and cadets who have checked out in event that we are in the path of high winds. . . .

I think it is high time you learned a wee bit about what I am doing so that you can tell people what "instruments" is—or are. First, aircraft instruments are not inaccurate, and when intelligently used are accurate to amazing degrees. A compass should not be off more than 1°, an altimeter 10', airspeed 2 knots, etc. Now as for the flying—there is of course little danger in a Link unless you get panicky and bail out. The parachute has not yet been devised which will open in five feet. All flying is dual, and the instructor is required to fly contact. No flights are made in instrument conditions—that's done only in R.A.F. training, not in the U.S. We don't wander all over the sky. Flying the beam is a very exact and exacting science. It is possible to orientate yourself in any radio range in about two minutes with no visual sight whatsoever. From there on all you do is assume a given heading, intercept and ride your beam, make your letdown, and the exact time over the field can be determined. In fact, must be determined, that is the whole idea of this phase of training, i.e., exact, smooth flying. And if I get out of here to Advanced, I'll be just that, for the check pilots are the worst anyone has run into so far.


August 25, 1944

Dear Phyl,

Not much to write, or time to write. Things are beginning to pile up. Sunday will be a big day with a check ride in the Link and finals in Navigation and Engines, tests in code and blinker. Still no plane rides, but I'm hoping every day that "tomorrow" will be the day. I have really enjoyed the Link hops and wish it was not ending. The Link Building and Ship's Service are about the only decent places on the base. . . .


August 27, 1944

Dear Phyl,

. . . This was quite a day. The finals in Navigation and Engines were rough, and I'll be happy when I learn I passed both. The Link check was satisfactory to the instructor, although I suffered quite a case of checkitis. No flight today (that's getting old, isn't it?) and as I have watch tomorrow, Tuesday should see me starting the old grind again. As it's been six weeks since I last flew, I should do a brilliant job of coordinating.

August 28, 1944

Dear Folks,

This afternoon I am on Messenger watch in the Batt office, and having everything under control, am getting off a few overdue letters. Practically all the news around here took place yesterday. First we had our Navigation final, which I passed, and ended with a 3.4 average. Not as good as I would have liked, but there is nothing to make one eager to study, and the instructors make it as unpleasant as possible.

Second came an 80 question engines test, which I understand everyone passed, I hope. It covered a three week review of engines and asked extremely detailed questions. Frankly, I could never make a mech, it's interesting, all right, but I can't quite see it. Everyone is thoroughly disgusted with the course in Engines—practically no stuff a pilot could use, reams of theory and construction, a dozen meticulous dull movies from which at one sitting we are supposed to gain the finer points of operating a Hydromatic prop—and then an 80 question test over everything that had not been stressed.

I was quite surprised, the other day, to hear a flight officer admit to a lecture group that cadet morale was the worst ever known in Naval Aviation history; and he laid no blame on us, nor had any cure to suggest.

Third event of yesterday was my Link check, which was satisfactory. I suffered an amusing and upsetting case of checkitis, but managed to come thru where it counted. We have a standard procedure to find the right heading to use on a beam to correct for drift, and I had an awful time using this—bracketing—on a beam under no-wind conditions.

So tomorrow I hope to start flying. The last two weeks of ground school are spent on gunnery; the first week is theory, the second week is range work with .30 and .50 cal. B.A.M.'s [Browning Automatic Machine guns]. . . .

SNJ North American trainer flown in Instrument and Advanced training
squadrons at Corpus Christi (Courtesy U.S. Naval Institute, Library and
Photographic Services)

August 31, 1944

Dear Folks,

This has been a pretty good day, mainly because I finally got my
first hop. With it returned any lost morale (and I assure you *that* has
never bothered me since Preflight) and all of my enthusiasm for flying.
The J is a wonderful plane, comparable to a Buick or LaSalle in cars.
Power to waste—they take off at a little less than half throttle, retractible
landing gear, hydraulic flaps, constant pitch prop, and sensitive controls. No more "kicking" rudders, just a slight pressure for a 45° bank.
We cruise at 130–140, and they'll do 210 straight and level at top speed.
They fly "by instrument" much easier than a Link and respond exactly
to control pressures. As a comparison, in the Vultee, I had to stretch
my leg out completely to apply full rudder, and in a J it takes a little
ankle pressure and a slight straightening of the knee. I can imagine
acrobatics in them would really be swell. To top it all off, I have a swell
instructor, even better than the one in Basic—a little fellow, not as big

as I am, and we get along fine. As ground school is practically over, I can really concentrate on flying. A word about the J; it is the SNJ-5: S for scout, N for trainer, J for built by North American, who also builds the B-25 and the P-47 Thunderbolt. The J is the best known training plane in the Navy, is used for actual scouting in patrol work, can mount two .30 cal. machine guns, one firing through the prop, the other in the rear cockpit. It is tricky to land, and I've seen some awful landings here, despite the fact that the instructors make all the landings. My boy isn't too sharp, but I figure he can do as well as I would for a while, anyway.

Dad's letter on the affairs of state was much appreciated, and I am in general agreement with all points, I believe. I certainly agree that should Roosevelt get a 4th term and not live it out, Harding would look like a national hero. This general opinion is also shared by Missourians who know something about Truman's home politics. Also 4 terms is carrying things to an extreme, and the good points of the Roosevelt regime seem to be becoming fewer as the years pass. . . .

. . . September 1, 1944

The big day has come and gone, and we are all still in the Navy despite the best of scuttlebutt. Had my second flight today and am fully as enthusiastic as yesterday.

Did I tell you that one of my Liberty roommates, W. T. R—— of St. Joseph, Mo., is here now. Our old room and platoon have done well. Of the nine of us in 215 Melrose, three are here at Chase Field, four are at Pensacola, and one washed, and one quit.

I am enclosing the record of my Link time, to be put with my W.T.S. log books, if you can find them. I have no idea where they are unless in the right hand drawer of the desk in the den or someplace in my dresser. Anyway, keep this record as the time isn't logged and 15 hours of Link time for a civilian would cost several hundred dollars. Also find enclosed the remnants of the last two weeks' pay.

Had a letter from Bill Roy the other day, among other things urging me to get into multi-engined planes. Oh, well, he may be glad to see me doing barrel rolls around his PV some day, I hope. . . .

Talk about rain: Wednesday (liberty) coming up from Corpus we ran into a terrific thunderstorm. In ten minutes there was water clear across the road. Clouds were about 150'–200' high and pretty nearly black.

This talk of "something happening" isn't all scuttlebutt. This squadron will be the only instrument squadron soon, as Main Base squadron is closing. Cuddihy is now fully pre-operational, and four admirals are coming to confer on aviation training. . . .

September 1, 1944

Dear Phyl,

. . . I now have four flights under my belt and am feeling good about instrument flight. My instructor is a swell little guy who never bitches about what I do, and to listen to him you'd think I was an H.P. [Hot Pilot] on partial panel work. Everything is "Swell, Rea, that's the way I want it, wonderful!" Well, this boy is used to a different technique on the part of instructors and a great many unprintable comments on my flying, so the change is pleasant. At least I feel that my instructor is sincere in what he says, so I must be getting on satisfactorily at least. Which is something around here, believe me.

The SNJ is a wonderful plane; every day I grow to like it more and more. Plenty of power and speed and as delicate on the controls as anything imaginable. You really fly it with three fingers. Today coming back to the field the airspeed hit 200 m.p.h., and that's a mighty nice speed to be cruising along.

We have few accidents in the squadron, five fatalities in the last year, compared to about twenty-five at Pensacola. Most of the trouble is instructors getting lost (one yesterday was gone two and three-quarters hours over his flight time) and a few ground-loops now and then— the only danger being fire, and that slight. Most of our flying is done between 6,000′–10,000′, so there's quite a lot of sky for everyone.

Have the big gunnery final this afternoon, so I'd better get on the ball and put in a little time studying muzzle velocities, bombing angles, and the weight of the MK12-Mod 2-AN-MK3, which means about as much to me as it does to you.

September 5, 1944

Dear Phyl,

I don't see how they do it, but every day they find something more for us to do or some new way to make life unpleasant. This week they (the officers-in-charge) decided to change the liberty schedule and began the new order by cancelling this week's liberty; so instead of having all day to write you, I have half an hour between breakfast and muster. Everybody is quite happy. By the new schedule we get one overnight in eight days and are secured from flight a half day, instead of having a whole day off. As this doesn't give us time to get much farther than Beeville, it pretty well cancels our liberties.

This week in ground training we are having practical gunnery. This includes firing the .22 pistol, shotgun, and .30 cal. machine gun, and field stripping the latter. I can get one of the darned things apart, but putting it back together is another problem.

I've had half of my flights now and am rounding the home stretch toward the old check ride. Three of my roommates had radio checks yesterday; two got ups and one a down, which is a pretty good average. . . .

September 6, 1944

Dear Folks,

Not much in the way of news these days. Flying goes on as scheduled. I have four more flights before my first check. I have a lot to learn before then, so am making no prophecies as to the outcome. The check includes the following things: steep turns at 30°, 45°, and a "C" pattern which takes a lot of hot flying and looks like this on paper—

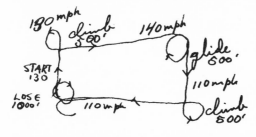

The trick is to time your turns to come out at 1½ m. and 2½ min., hold your heading during changes of speed, and keep relaxed, with your wits about you all the time. After this you cage your gyro instruments, and the check pilot puts the plane in some violent dive or climb, turning and usually skidding or slipping. The last thing is the "practical problem" which starts with an unusual position, and then you turn to a given heading using clock and magnetic compass, reach a given altitude, fly straight and level for 2 minutes, then make a 180° turn and begin a let down at 500 ft/min. and 110 m.p.h. None of this is difficult to grasp, and none of it actually difficult to perform; the trick is getting your brain and the plane under control in the short time given to you. It doesn't take flying ability, just mental ability—and nobody laughs at the saying that you lose 40% of your mentality under the hood, least of all, me. Three of my roommates have finished, two on extra time, and one is even up on his third ride tomorrow. Some of our class are ready for their radio check, others are waiting for extra time on their P checks. I'd better explain that the first check is the "P" check (for primary instrument work), the second is "R" check (radio range work). This morning we had a lot of low clouds and rain, so no flight, and as I have liberty Saturday, I should check Monday. Six fellows got the ax from the board today, but some are getting through and I intend to be one.

Speaking of liberty—we are all very much disgusted with it. The O-in-C of the Cadet Battalion was retired the first of the week, and the new fellow is a very unpleasant character who is too eager for the good of our military careers. The first thing the new O-in-C did was to cancel our liberty this week and establish a new liberty system. Instead of having Monday night and all day Tuesday off, we now get one overnight in eight days with half a day off, missing flight but not ground school. And if you have afternoon flight, you don't get the overnight. This means San Antonio and Corpus Christi are next to impossible to make, Beeville is a sad town for liberty, so there's little to do but get in a little extra sack time.

Gunnery this week is very interesting. I shot a 4.0 with pistol at 25 yards, hit 13/25 at trap, 17/25 in skeet, and did fairly well with the .30 cal. machine gun. Skeet and trap are rough, because we are using a simulated reflector sight instead of the usual off-hand shooting. The sight looks like this:
and you have to give the right lead by rings and according to flight path. Same idea as fixed gunnery in a plane. Trying to learn the mechanics of the machine gun is rough. The instructors know them forward and backward blindfolded and expect us to pick it up without any trouble. Today we belted ammunition for an hour—practical training, I guess. . . .

September 6, 1944

Dear Phyl,

. . . This, if I get it done, will be a masterpiece of concentration as my roommates are talking both flying and women, which isn't conducive to good letter-writing.

Things are going on schedule around here. I have four more hops before my "P" check and still have a lot to learn. Six fellows got the ax today, but three of my roommates have made it through, so I see no reason why I shouldn't. We had a heavy overcast and rain today, so I missed a hop. Liberty comes Saturday, so I'll probably check next Monday.

We're having a lot of fun shooting in gunnery. So far I've been hot with a pistol, and today had a good run on the skeet range.

The entire reading population of the base is eye-deep in a new novel written by a Navy man concerning the activities of the Naval Air Corps on liberty. It is entitled "Shore Leave," by Frederick Wakeman.* While it's a pretty rugged little number, and definitely not on the approved reading list of a JuCo for young ladies from Maine, I think you might find in it a pretty real picture of the way my buddies live and think. . . .

September 7

Another day, almost as bad as yesterday. No rain, but low ceiling and no flight. Consequently we sit around and get in a lot of hangar time. Subjects of conversation include flat-hatting and the usual results—crack-up or getting busted, cross-country flying and good liberty towns, how to make parachute jumps and when, the latest scuttlebutt, washouts, how much an ensign can get away with, and so on far into the morning.

We had the big talk today about 'chutes. Here the first guy to fly takes his chute out to the plane and everyone else uses it all day. This means that a small fellow may get a chute that won't even stay on him sitting, and much less jump-worthy than an umbrella. The idea of going with a plane doesn't appeal to most of us, but that would be better than trying to jump in a large chute.† At least it gives us something to talk and argue about.

We have three hours on the range today, so it ought to be an easy afternoon.

---

* New York: Farrar & Rinehart, 1944.

† This argument proved decisive at a later date when Rea was involved in a midair collision.

September 8, 1944

Dear Phyl,

Another day, cool, cloudy, and promising rain, flying conditions undesirable to bad, so we are grounded again. It would appear that the rainy season has set in in Texas. Funny thing about Beeville, yesterday we saw heavy rains all around the base, but not one drop fell here to secure flight or P.T. The rains did cool us off, though, and last night was a wonderful night to sleep. . . .

Had a great day on the range yesterday. Got hot on the pistol and hit 4.0, had a good run on machine gun, and after a lousy start on skeet, hit 14 of my last 15 for a 3.5. These sights we are using, simulated reflector sights, really are the gouge to shooting a shot-gun. Once you get the range, you can't miss.

Some of the fellows are buying uniforms now—blues, greens, and greys. The cost of dressing an ensign is amazing. A full set of uniforms runs up to $260, which is a lot of money for anyone's clothing. It isn't too bad, though, as we get $150 clothing allotment upon graduation.

The nice part about ordering now is that if anything happens and you don't need the glamor-rags, you don't have to pay for them. Graduating in the fall means a little more outlay in blues and greens instead of converting whites and greys. I sure hope I'm wearing a set of greens the first of the year and not a pea-coat and flat-hat.

September 9, 1944

Dear Phyl,

I am celebrating liberty after a week and a half of sitting around the base by writing to you. Last night I investigated the potentialities of Beeville, and hurried back to the base. What a sad town—even Olathe, though much smaller, was a nicer place than Beeville.

. . . This morning flight was again secured because of a hurricane warning, and preparations have been made for taking the planes up to Dallas, just in case. No one expects anything to come of it, and we have higher ceiling today than any time in the past three days. That's the Navy for you, though. . . .

September 10, 1944

Dear Folks,

If the weather in Southern Kansas is like that of Southern Texas this Sunday afternoon, Dad is either missing a good chance or out catching up on his fishing. After a week of pretty foul weather, today has come through with a bright sun that is hot in spots but leaves the shade nice and cool.

My instructor has liberty today, so I am missing my fourth or fifth consecutive day. However I have reached the point where sitting around doing nothing doesn't bother me any more. Now that ground school is over there will be nothing to do for a half a day, so I can concentrate on flying and catch up with my letter writing.

Gunnery came out quite satisfactorily. I made a 3.4 on a pretty stiff exam on .30 and .50 cal. machine guns and had a good day on the trap range, breaking 20 of 25 birds—8 is passing, when using the sight, and plenty of fellows sweat out that low score. My best run was on skeet, where I broke 14 out of 15 birds.

Three of my roommates went back to the main base for Advanced pool today, so I have some new ones to break in, now being the "old man" of the room. My Wichita buddy P—— also went in today. In the new group is another of my old buddies and former roommates, Bob P——, so I shall have to give him the word on Beeville sometime soon.

We are plagued with eagerness at present. Our old Batt officer left, and the new one is hot on military discipline. To aid him four new Marine 2nd Louies came in and are making themselves unpopular as fast as possible. I probably told you about the new liberty deal, whereby we lose substantially, the imposition of a 50 mile travel limit putting San Antonio out of bounds. The latest deal is a cracking down on communications. Heretofore we've passed 12 w.p.m. in Code and been secured from daily practice. I've been getting 14 w.p.m. without too much effort and no practice. Well, now we have to get 16 w.p.m. to secure from practice, and the code is lousy, jumbled, and ill-timed. About 3 in 400 cadets are getting 16 w.p.m., so we all have 30 min. practice each day. I'm hoping to get up to the requirement, as I know I can take 16 in good code and was taking 18 w.p.m. at Olathe. . . .

Lots of talk going around now about buying uniforms. It's imperative that they be ordered soon if you want them for graduation. If and when I finish my work here, I shall investigate. One of my buddies filled his needs with an order of $260, but I can't see that much when the

Navy only allows you $150. We'll have to get greens, greys, convert blues, and a hat, at a bare minimum—and believe me, they soak you plenty. Working greys—same material as my cadet khakis—cost only $12 (coat and pants), but greens cost $70–$90 even without the necessary accoutrements which adds about $10 to any uniform. With a $15 hat and converting cadet blues for $12, plus odds and ends like collar pins, wings, hat covers, it doesn't take much to run up a good bill. At least you can order and not pay for them if you don't need them. . . .

September 13, 1944

Dear Phyl,

. . . I'm having a lazy life around here. Sleep until 0900 or 1000, eat whatever I want for breakfast, no studies, only one flight a day. It's not as nice as it sounds, however. I am really sweating out my flying and will check day after tomorrow. I know how to do the stuff, and I know I can; it's just a matter of keeping everything under control and doing it. My lay-off of five days last week didn't help, but yesterday's flight was somewhat encouraging. If I get an up on the first ride, you'll get one of the happiest letters from me you've ever had. . . . A lot of my buddies have gotten downs, two of them are on extra time now, so there's no telling what will happen, but we'll hope and work for the best.

September 15, 1944

Dear Folks,

For once I have something to write about besides the Navy, so here goes. Wednesday evening, feeling rather fed up with everything and having 28 hours of liberty on my hands, I gathered together my few remaining shekels and boarded a bus for San Antonio. I wasn't quite sure whether the trip would be worth the expense, but I find I am back with cash still in my pockets, a swell liberty back of me, and high morale, so I'm high in praising S.A.

Got in late, 2330, and my luck began at the second hotel I visited in search of a room. There I ran into an AAF corporal who was lined up with a double room and in need of another person. It turned out he had been an AP [enlisted pilot] in the Navy and flown off the Sara[toga] and Lex[ington] (so he claimed) and wouldn't hear of me splitting the bill on cadet's pay. So, for free, I had practically a suite in a good hotel.

Figuring there'd be a lot to see, I rose early, breakfasted well, and set about seeing the city—and I do mean city, not town. Incidentally, that breakfast was the only good meal I had in S.A. They seem to be short on good eating places or else I've lost my ability to find them.

After the business section, I found my way to the Alamo, expecting much and being somewhat disappointed in what I found. At a USO I picked up a "Soldier's Guide to S.A." and so my next stop was supposed to be the 200 yr. old Mexican settlement called La Villita. However, I found that the Red Cross got there before I did, so that was out. So, next I hopped a bus and after a little walking found myself out in the country at Mission Concepcion—according to my Guide Book—founded in 1731. This was much better than the Alamo and not rebuilt. It was a pretty good-sized mission and must have been rather beautiful before cracking plaster, bullet holes, sun and rain, and tourists' inscriptions faded the painting and decoration. I spent an interesting hour out there with the guide, a little dried-up old Texan who seemed to enjoy the visit as much as I did.

It was noon by now, so back to town and dinner at the Plaza Hotel, which was quite a let down. Beautiful dining room—poor food. Had a rather interesting table partner, however, a Dallas man connected with Interstate Bus Lines and newly stuck with the public relations end of a truck line acquired by his company. Apparently the truck line business is still what it used to be, plus a few new headaches in the form of women billing and rate clerks and checkers.

After lunch I visited the old Spanish governor's "mansion." Here the Daughters of Texas Revolution hold sway and it was in good shape, with a beautiful garden patio and, quite typically, a couple of handsome senoritas in bright red and green dresses to add a touch of life. For contrast, I visited the old San Fernando Cathedral on the Plaza and the Buckhorn Curio Shop. The latter was quite a place. I never saw so many heads and horns in one building. Specialty was steer horns and they had every imaginable size and variety.

Having done enough hiking for one day, I went down to what is called the S.A. River, a sort of glorified irrigation ditch built by Indians and later landscaped by W.P.A., and paddled its length in a canoe. It's an amazing little park to find in the heart of a city and quite delightful for an hour of cool relaxation. It's about as straight as a corkscrew, and any town but S.A. would have filled it up before now.

Among other things in S.A. there's the Army. Never saw so many in one town before. Air Corps cadets are even thicker than we used to be in K.C., but they sure aren't the same fellows—they either look like kids or are walking around with their wives and babies in tow. . . .

Mother asked in her last letter how the visit of the 4 Admirals came out. Well, it was not for the best as far as I can see. We are informed that we are bad boys and not military enough (when was the naval air corps that?) and so we'd better be good or we'll get put on report, and if we are good and march in step and don't talk in ranks and keep the upper layer of Texas out of our rooms, we will get chits that will entitle us to special liberty, providing we don't have any extra duty to walk off. If we're excellent, we get 2 chits, if we're good, 1 chit, if we're average nothing is said, if we're bad—5 & 1 [five demerits and one hour of marching]. And in order to see that we stay approximately near, if not on, the ball, we have 3 boot 2nd Lieut.'s USMCR to herd us around. Boy, what some of these disciplinarians don't know about handling men would fill a large volume. . . .

September 16, 1944

Dear Folks,

Not much news today, but what there is is good, very good. I had my "P" check today and, despite a rough day, flew a good ride and got a good up. It was one of those days when you just can't miss, and I flew a good close C pattern, smooth recoveries from unusual positions, and a well-controlled practical problem. I had a good boy for a check pilot, and he kept me at ease and set any doubts at ease by doing a slow roll immediately after he popped my hood. They don't do that if they aren't in a good mood. Consequently I am feeling pretty good this evening— right on top of the world.

Tomorrow morning, I'll start on my radio hops—only three of them, and then another check. I'm not greatly worried about this one (I'll admit I was sweating out the P check) because I know my radio work and now have confidence in my ability to fly the plane by instruments. If all goes well, I'll check out of squadron on Tuesday or Wednesday and have time to visit San Antonio and Corpus again. I hear the pool for Advanced at the main base is from 5–15 days in length, so I'll likely have another long vacation from flying. But when I do get to see where I'm going, this boy is really going to open up that throttle and watch the sky and clouds go by. Today we flew at 10,000', my check pilot peeled off at 9,000' and came down in a 210 m.p.h. dive that really brought back the thrill of flying. . . .


September 17, 1944


Dear Phyl,
    . . . Today I had my first radio hop, and I only hope that the rest of them go as well. It was a swell ride, and my instructor (a new one for each R flight) said it was the smoothest 1R he had seen. Sweet music to my ears, believe me.


San Antonio, Texas
September 20, 1944


Dear Folks,
    Just a note to let you know that things are turning out better every day. Yesterday I had my Radio check and got the big up-check, so one more step is behind, and one and only one more lies ahead. It's a bit hard for me to realize that I am actually in Advanced and that if and when I finish the next base this cadet life will be finished and it will be Ensign and not Kaydet.
    I really had to sweat out my check. There were seven of us ready to check on Tuesday, and we all expected to be scheduled for the first morning flight, when the air is quite smooth. To our disgust, we found we were all scheduled last period, when it starts getting pretty rough. Added to this, they gave us a group of new check-pilots whom no one knew. So from 0715 to 1145 we sat around and watched the cumulus clouds build up. However, we were either a good group or very lucky, for five of us are in San Antonio today. We have liberty for the rest of the week, so far as we can stretch our finances.
    I got a note from Class 6 [Mother's Sunday School class], and will answer it promptly. Anytime T—— wants my fencing mask, he is welcome to it and any of that gear. As for my birthday, have a big dinner for me, but I really don't need a thing. Two ideas hit me—I might call

home, and cash is always a handy commodity, especially when you have to pay ⅓ down on uniforms before you get your allotment. I'm not sure how to handle the prospects of graduation, 'cause I don't like the old Navy custom of going way into debt as soon as you are commissioned and then using the next six months' flight pay to get yourself out. Tomorrow or Friday I'm going down to Corpus and look into the uniform situation. I think I'll get as little as possible here and wait till I can get back into the continental U.S., where they don't raise the prices quite so much, to buy the rest.

We will go back to the main station on Sunday for 5–15 days of Advanced pool and then on to Kingsville or Waldron for Advanced. The final phase will last 2–4 months, but I am hoping to be home on Christmas or thereabouts. Idle dream, but there's always a chance for a break. They're giving 15 day graduation leaves now, and that wouldn't be hard to take. . . .

Main Base, Corpus Christi
September 25, 1944

Dear Folks,

Not much to write about. This is a pretty soft life. We have one hour of P.T. and one hour of code and that's all for the day. Reveille is at 0700. It's a nice change, but a week of it will get pretty monotonous, I think. It does seem pretty good to be back here at Main Base, though, after six long weeks at Beeville.

I think I've found the gouge on uniforms. Went over to the Ship's Service this morning and looked at their uniforms. Nothing fancy, but good material and quality and prices. Regular Navy quality uniforms at exactly one-half the price of any of the downtown dealers. I can get greens, greys, convert my blues, and an aviation grey overcoat for less than $150. Looks pretty good to me. While my clothes won't have the name of some New York tailor in them, they will be the same clothes and about twice as many of them. So I may be wanting a little money shortly to pay the 25% retaining charge until the clothing allotment comes through.

We had quite a little rain storm today and it is actually cool for

once. It's beginning to show signs of a change of season down here—
getting dark about 1930.

Nothing else to write about. If you want to see some fine pictures
of naval aviation and articles concerning the same, get the last copy of
"Flying" magazine.* It's well worth 50¢. . . .

September 27, 1944

Dear Folks,

My mail caught up with me yesterday, and to say I was pleased,
surprised, or overwhelmed with what I found would be a masterpiece
of understatement. I don't know quite how to thank you, but believe
me, I do, and will see how far and how much I can make that check
cover. A better birthday present could not have been found.

As I wrote you in my last letter, I've found that Ship's Service is the
best place to buy uniforms, so I don't think I'll be going in over my
pocketbook. And I'll look around carefully for a watch, too.

We are certainly seeing another face of Texas this week. For three
successive days it has rained, lightly, heavily, intermittently, and steadily. Yesterday afternoon one of my buddies and I started back from the
show in a slight drizzle, and when we reached the barracks we had to
pour the water out of our shoes. We have raincoats that keep the top
half dry, but they are of a rubberized material that makes shirts and
pants look as if you'd been working in a coal mine.

Besides limiting our activities around the base, this current bad
weather means just so many more days in the pool, for no one is flying.
We can get refresher hops out at Cuddihy if the weather ever clears up.
Advanced squadron is supposed to last seven weeks, I hear, with the
length of time depending quite a bit on which squadron you're in. There
are squadrons here at Main Base, at Waldron, and the largest at
Kingsville.

Thanks again for the swell birthday, and also for the word on local
candidates. I'm getting my ballot off this afternoon. Will write as things
develop. Today we got our blue name-plates, indicating that we are
upper-classmen.

*Vol. 35, no. 4 (October 1944).

September 30, 1944

Dear Phyl,

. . . Yesterday I got in a "gravy hop," that is an extra flight, sort of on the cuff. It was the first time I had really flown for eleven weeks, and it was pretty sad. We tried to fly a little formation, but that was so bad we decided to shoot landings—less dangerous. It was fun, though, and I'm glad I got it in before going to Advanced. I'm going to try for one or two more before I leave, and see if I can't get back to where I was when we finished Basic. I'm fully expecting to be here another full week, so I should have plenty of time.

Not being able to get a flight this afternoon, although we went out to Cuddihy for that purpose, I settled for an hour's sailing. . . .

If all went as it should, I will (would) graduate about the 15 December. However, I think there is but a very slight chance of finishing before the first of the year and will consider myself lucky to be thru then.

October 1, 1944

Dear Folks,

Sunday afternoon and I'm listening to the New York Philharmonic, so it seems an appropriate time to write home. Day before yesterday, nine of us went out to Cuddihy for refresher hops. Pretty rough flying after an eleven week lay-off. So yesterday we all went back for another hop, only to find out after we got out that no planes were available. Being some little bit put out at this, I decided to buy myself a birthday present I've been wanting for the past year and invested in a good pair of sun-glasses at the Ship's Service store. Returning to town with a few extra hours to kill, I decided to try the sailing for the day and spent a delightful hour out on the bay.

Continued Monday—Sunday afternoon we had a big regimental review. After 16 months, this sort of thing leaves me pretty cold. You stand around for an hour, then march around the quadrangle forming a regimental front. Much presenting of arms, attention, and parade rest as the admiral drives up. Then the cadet officers do a "front and center," the band marches up and down for five minutes, the heat gets a few cadets who pass out, the rear ranks are in a very unmilitary state with ties loosened, caps askew and rifles at anything but parade rest, while the front ranks look very sharp; a couple of Marine aviators are awarded the DFC and Air Medal, more presenting of arms, the band plays the National Anthem in waltz time, we "pass in review" and march off, hot and sweaty; one company gets an extra hour of drill for its unmilitary bearing when under arms; we stow our toy rifles and make a beeline for our rooms, shedding hats, shirts and pants as we go, wondering if there is a more unmilitary outfit in the U.S. armed forces, and not caring much. That is the Navy air corps—lots of esprit d'corps, but only when someone questions it.

I spent a very quiet birthday. The usual PT and code in the morning. In the afternoon five of us went to town, a show, and dinner, returning early. I thought I would call home, but decided to wait until I got my orders for Advanced, which should be in three or four days at most. I have another sailing party lined up for Wednesday. That strikes me as the most interesting pastime down here and certainly one I can't indulge in everywhere. . . .

We are all watching the days pass and the warm weather remain with a certain amount of dread, because on the 15 Oct. the uniform of the day changes to dress blues, which means no more liberties in sloppy, comfortable khakis, but white shirts and wool coats, regardless of temperatures. Officers will wear light-weight greys, but we will literally "sweat it out." Ran into another of my old Liberty buddies from Wichita the other day. It becomes more and more enjoyable to find a Wichitan or Kansan as the months and friends pass. . . . Don't worry about the quality of the uniforms I'm going to buy. There is some difference between $45 and $95 uniforms, granted, but I figure that most of the difference lines the pocket of one Nate Levine who sells the latter—and I like the Navy's uniform better anyway.

October 4, 1944

Dear Folks,

Just returned from an enjoyable and restful ninety minutes in the music room listening to the Shostakovitch "5th Symphony." At times it's almost possible to forget you're in the Navy. And at present, that would be a pleasure. It seems we have been bad boys lately. A Marine captain, upon seeing us proceeding to code class, commented that we looked worse than the Chinese Army. This, he thought, might be remedied thru practice, so our company is marching from 13–1400 every day this week. On top of this, a surprise room inspection caught us with our brass unpolished, so that added another hour of walking. Of course we drew drill for P.T. today, so after three hours today, my lower extremities are ready for a night's rest. . . .

I finally got my 16 w.p.m. in code this week, and 10 w.p.m. in semaphore, but doped off in blinker for the first time down here and only got 8 w.p.m. in blinker, so still have to go to class. I ought to be able to make it easier for myself in Advanced by securing code after the first week or so.

Speaking of Advanced, I finally "made the list" of outgoing cadets, which means I might leave in a couple of days—or a week. All is very indefinite, but I should know in a day or two. . . .

The P.A. system informs me it's time to make my bunk, torn up for airing day, as taps are 15 minutes away, so that's all for today.

October 5, 1944

Dear Phyl,

. . . The other day, Monday to be exact, our company was seen going to code by a Marine captain who was so impressed by our military bearing that he immediately called our Batt office and informed the officers there, whose dislike for us is only exceeded by ours for them, that we marched worse than the Chinese army, and suggested we could use a little practice. Consequently, the 23rd company has been marching around the quadrangle from 1300–1400 every day and is restricted from refresher hops for the week. Everyone writes about how nice and cool it is in Kansas, Indiana, and Maine, but Texas is as hot in October as it was in July. Also we had a surprise inspection on Tuesday and everyone drew 5 demerits and an hour's marching for one reason or another. Yesterday we had drill for P.T., and this, plus the hour for the company and my one hour extra duty—well, I'm buying bigger shoes next time I buy. . . .

This morning I went over to the Regimental Headquarters for orders to Advanced, but although I was not last on the list for orders, I was too far down to get them today. Tomorrow I should make the grade. A lot of the fellows are waiting around for sugary orders to easy squadrons, but this cadet is taking whatever comes along and making the most of it. Finally got the word on Pre-Operational. We'll have 3 weeks of this at Cuddihy Field flying SBD's for 9 hours and some 60 hours of ground school, then will graduate. This means that there's no chance of graduating by Christmas, probably by the end of January. . . .

October 6—Finally got my orders today, and will leave at 1000 tomorrow for Squadron 14A at Kingsville. I hear it's a fast squadron and comparatively easy to get thru if you don't crack up. This sounds good to me. I could have had Main Station, which is a cinch, but is very slow, and all my buddies will be out at Kingsville. Tell you more about it after I get out there tomorrow. It's about 50 miles west to Kingsville and pretty much away from everything.

Kingsville, Tex.
October 7, 1944

Dear Phyl,

So here I am at Kingsville, very tired as a result of getting checked in this afternoon. We thought main station was pretty well spread out, but it had nothing on this place. You have to take a bus to go to the flight line, and it's a good half mile from the barracks to the Ship's Service.

It's a pretty good looking base, well kept, with one-story barracks. The recreation building is even superior to that at Olathe and has everything one could want. There are hundreds of J's sitting out on the line, and with their red, yellow, blue, black, and green paint jobs, look like some sort of Christmas decorations.

We came in the rain, so there was no flying this afternoon, but we had quite a show on the way up from Corpus. The road passes near a field used for strafing runs, and we watched the boys come in at about ten feet and then pull up into slow rolls, snap rolls, and Immelmans at 200'–300'. We also learned that a couple of fellows got their wings the hard way in a mid-air collision last night. This isn't the safest type of flying in the world, and a certain amount of fatal accidents is to be expected.

However, we will not be flying for a week or two. Our flight of six men is in one room and so we eat, sleep, and fly together. I previously knew none of them, but they seem to be an average group and are doubtless as good or better than I am. I've found a large number of

my old buddies and have had quite a time "getting the word" on Kingsville. . . .

Sunday morning—What a set-up! We get up for chow at 0600 and
are back in the sack by 0630. It's now noon, and we are just getting
around to seeing what goes on today.

October 10, 1944

Dear Folks,

. . . I suppose the first thing is to cover the base here at Kingsville
and our current activities. This (it seems I'm always saying it) is the
most spread-out base yet. There are regular buses between the various
parts of the base. There are two fields for the four squadrons based
here. The Rec building is a beauty—even bigger and better than Olathe's
was, with almost everything anyone could need or want.

Cadet barracks are one-story frame affairs and look better than
most of their predecessors on the outside. The inside is the same as
everywhere else. Bunks are good, though. The six men in our flight live
in the same room, which makes it nice. All are quite congenial, and we
should get along well enough. Practically all my old buddies are here
too, although I don't see them very often.

We should go down to squadron (start flying) the end of this or
beginning of next week. This week in P.T. I am playing handball. I like
the sport tremendously but can't brag about my ability. There also
seems to be some question as to which will give out first—the week or
my right hand.

Breakfast muster is about 0600, P.T. muster at 1100. In between,
we sleep three or four hours or read. Very restful. The afternoons and
evenings are free at present, and I have investigated the library and am
learning the finer points of the game of bridge—slowly.

This evening I bought a khaki blouse from a roommate. I've wanted
one since Flight Prep, missed out on the issue, and as I only had to pay
$3 for a good twill jacket, am quite happy about the whole thing. Retail,
same coat costs $15. . . .

The weather here at Kingsville is beautiful at present. From about
1000–1800 it's warm to hot. The mornings and evenings are delightful,
and the nights cold enough for a blanket—the best kind of sleeping. . . .

October 10, 1944

Dear Phyl,

Here we sit around the one table with a light on it—all six of us.
Writing letters—Rea, reading—V—— and H——, talking about postwar "What'll we do?"—McG——, K——, and P——. My roommates.
The trick is for Rea to stay out of the conversation and concentrate on
the letter. . . .

There's little to write about here. We are doing nothing except take
P.T. an hour a day. At present I'm indulging in handball—great sport,
but it's a serious question as to which will give out first, my hand or the
week I'll be playing.

In the afternoon we play bridge. This sounds horrible, but it's true.
Today we played from 1:30 to 7:30. I'm learning the rudiments against
a couple of old hands at the game with reasonable success. Of course
my partner and I make some sad mistakes, but all in all, it's an enjoyable
way to spend the afternoon.

There is a pretty nice little library here, and the first thing I ran
across was a collection of horror stories. The choice one was "Our
Feathered Friends" by Phillip McDonald. Good one for your English
class when it gets mouldy.

October 15, 1944

Dear Folks,

Here I sit, resplendent in freshly pressed dress whites at eight
o'clock in the morning, wishing greatly that I was wearing blues. But
uniform regulations being what they are, the day's watch must shiver
a bit until the sun begins to give a little warmth. We are having beautiful
fall weather these days, and if there were any trees with leaves on them
within sight, I imagine they should be turning color one of these days.

This being the first day our flight has liberty since our arrival at
Kingsville, it is only fitting and proper that we should also have the
watch, a twenty-four hour duty stretch, which calls for the wearing of
whites that haven't been worn since I left Wichita. We were afraid it
would be too hot for blues, last week, but now, we'd be more than glad
to wear them. All isn't as bad as it sounds, however, and I think we will
be given liberty from secure watch tonight until 2115 tomorrow. I plan
on giving Kingsville the once-over and possibly going down to Corpus

tomorrow. As most of the men in this barracks are on liberty, there's practically nothing to do, so today should be as quiet as the rest of those I've spent here.

We are expecting to go to squadron on Tuesday and should get our first flight on Wednesday if planes and instructors are available. I am almost anxious to get at it again, although the syllabus here looks as hard as it does interesting. . . .

October 15, 1944

Dear Phyl,

Here I sit, lone and solitary, which is the way I like it when I'm writing to you. I'm on the old watch squad and am, at 0831, shivering in freshly starched dress whites, when I should much prefer to be in winter blues. However, the uniform of the day is just what it means, so, as I said six lines up, here I sit—and shiver.

It's a quiet day for watch, everyone in this barracks is on liberty, as I would be, conditions permitting. I think we'll have tomorrow off, though, so it won't be too bad. We are supposed to go over to the squadron on Tuesday for a day of lectures and should start flying on Wednesday. I am a wee bit anxious to get up again, although a mite dubious, also, about how it will go. I get quite amused at my non-flying friends who scoff so at my worries, great or otherwise, about my flying ability. Were they here, they'd soon see why I never write overly optimistic reports on my progress. Of course we write to our buddies at Pensacola and other places about what hot pilots we and they are, but there is a mutual understanding there that admits our faults and weaknesses.

October 17, 1944

Dear Phyl,

Today we got started. Spent the entire day at squadron—including noon chow period. Got the word on the SNJ, flight rules, fields, and all the things that seem unimportant but can really throw you for a loop if you're not on to them. Tomorrow we should get our first hop, and I'm really looking forward to it.

October 18, 1944

Dear Phyl,

... The screws are on and I need some moral support. Today was the big day for our solo checks, so I proceeded to fly a big fat down. I was a bit perturbed at first, but by now have talked myself into a good frame of mind and am more confident than I was when I went up this morning. It's nothing to worry about, I keep telling myself; even if I don't make it next time, I'm a cinch for [extra] time, and if I can't fly a J by then, well, better a live S/2C than a dead Ens.

October 20, 1944

Dear Folks,

Things being under control, I can report on the situation. First, the cookies were wonderful, and I am hoping for more in the future. However, I'm returning the can for refill as it's the only way to guarantee safe and unsmashed arrival of contents. You can put the stuff I'll send home in mothballs, I don't expect to be needing it any more.

Flying weather is still good and we are taking advantage of it, flying as much as possible. I didn't get the word on the J on my first try, got a down but proceeded to fly a pair of good ups, soloed today and passed the precision landing check. Things go fast here. The first ride is an area familiarization hop in the rear cockpit. The second one is a landing check of two hours. My boy was unable to get the word to me, and we both agreed a couple of rechecks would help. They did, and both my recheck pilots knew how to instruct and I got very good write-ups from both of them. Then came one solo hop and the precision landing check. All satisfactory. Last of my four flights today was an acrobatic checkout in an N-3-N. Tomorrow we start in on acrobatics in the J and check on them Monday. No flying on Sunday.

The J is supposed to be tricky to land and can be if you're not awake every second on the way down and while you're on the ground. But it's a sweet plane and will do anything you want it to. . . .

October 22, 1944

Dear Phyl,

   . . . Last week we ran into a couple of Texas A & I coeds who invited us in to an Open House at the Kingsville Methodist Church, to be a hayrack ride or some similar kind of party. We accepted, of course, and at 1930 last evening, four of us were knocking at the church door, not quite sure whether to walk in, or what. So we hear a sweet feminine voice saying, "Come on up." We give each other the old eye and dash for the stairs, up them, and burst in on the Open House. Then, and only then, we learned that this was "work night" and no party! There was no chance for a decent withdrawal, and so we sweat it out, stuffing bulletins, stapling church pamphlets, etc., until 2200. We had a pretty good time, after we got over the shock, and got a lot of work done. The amazing thing was that all four of us got up in time this morning to go back into town and to the same church! Now who says I'm not a good Christian youth! Hey! quiet.

We didn't fly Saturday, so we'll begin acrobatics Monday and probably be into formation work by the end of the week.

October 23, 1944

Dear Folks,

Boy, I mean I'm really tired tonight. This afternoon we had three hours of acrobatic flights and it was more than a sufficiency. Acrobatics in a J are easier than in a Stearman, but you have a lot of airplane to herd around, and arms and legs get plenty tired, not to mention a headache which usually accompanies an hour of inverted maneuvers.

We have only one instruction flight, three solos, and then our check, which is usually given with the check pilot on the ground. Our sequence of acrobatics is as follows: loop, left and right slow rolls, left Immelman, four wing-overs, right Immelman, two-turn precision spins left and right—all to be done in less than eight minutes. It keeps you plenty busy, but you can roll out of one manuever and right into another. The weather is perfect for flying, warm and usually clear. Today I flew in my skivvy shirt and summer flight jacket. Between 5:30 and 6:30, though, we had a lot of haze, and at 7,000 feet you couldn't see any horizon and the ground was barely visible. That's not a nice feeling when you're not familiar with landmarks. . . .

. . . Tomorrow we have our acrobatic check and then get started on formation—section and division tactics. . . .

October 24, 1944

Dear Phyl,

. . . Flying is going well but not too fast. Yesterday we had three acrobatic hops and were glad to get a fourth cancelled. Today the weather went back on us and we only had one instrument hop. Tomorrow we should have our last acrobatic solo and check ride. Acrobatics in these J's are swell, but you have a lot of airplane to herd around, and three hours of slow rolls and Immelman turns really wrings a guy out. As soon as we finish acrobatics we start in on formation work—section of 3 planes and division of six planes, with a multitude of different formations and break-ups. It ought to be fun and a lot of hard work.

October 26, 1944

Dear Phyl,

. . . Little has happened around here. Yesterday we had our acrobatic check. The check pilot stayed on the ground and couldn't see us, so we had to do it all over, but with the most satisfactory results.

Thunderstorms over the base kept us grounded today, but it's clear tonight so we'll surely fly tomorrow.

October 27, 1944

Dear Folks,

Pretty rough day, this. We had our first two formation hops and they left much to be desired. In fact, three of our boys got a down on the second one, and our instructor is none too happy with the prospects of our check tomorrow. I know I'm not too sharp at this kind of flying, but I think I can hold up my end of it. Section tactics require not only a lot of flying skill but plenty of headwork as well.

Now that Tex. is into its fall season, we are wearing blues for a liberty uniform—which means white shirts again—but will feel pretty good on cool evenings.

T—— wrote that he is having a short vacation from his med school grind and will probably drop over for a chat. Be sure to let him use any of my fencing equipment he may want. Fencers are still the same; he ran into a fellow at K.U. who claimed to be a pretty sharp swordsman and also held the State Championship. However when he found out who T——'s mentor was, he quickly changed the subject. Those were the good old days—how well we all now realize it.

About politics among the AvCads—We are probably pro-Dewey but feel sad when comparing the campaign speeches of Dewey and F.D.R.; the former sounds like a high-school debater at times, and I have yet to hear him put forth any concrete plans. However he specifically promised to "return the boys home speedily after the end of the war," so we are for him to a man. . . .

October 29, 1944

Dear Phyl,
. . . Yesterday three of us had our section tactics check and got an O.K. on it. I had the lead for most of the period and got a pretty good grade on the hop. I was a little worried about it because I'd been flying wing on all the previous hops, and one of my wingmen had gotten a "no good" on one of his hops and had never flown wing before. Nice combination, but it worked out beautifully. Now we go into six-plane division tactics, four-plane division, and fighter opposition.

October 30, 1944

Dear Folks,
Today we had our first six-plane division hop. It was great fun for me because I flew as passenger with one of the instructors and had nothing to do but watch. Incidentally, if I hadn't mentioned it, we got an "up" on the section check on Saturday afternoon.

This division work is really interesting. Our syllabus includes division dives, six-plane echelons, columns, and a division attack. The whole secret of this kind of flying is teamwork and good headwork.

You asked about the instructors here at Advanced. We have all kinds—some of them just out of instructor's school, others who have been here at Kingsville for two years, and still others who have spent their eighteen months with the fleet. I had a fellow for an instrument hop who flew TBF's off Guadalcanal about the time Bill Roy was driving his P-boat around Rabaul.

Ground school is pretty haphazard and rather easy. I've already checked out in code at 16 w.p.m. and semaphore at 10 w.p.m. and have really gotten on the ball in navigation—which makes me feel good. Our last nav problem I finished ahead of the rest of our class and got a good 4.0 on it.

Did I mention the Link nav hop we had? It was pretty wild trying to navigate and fly a Link at the same time, plus changing course every four to seven minutes, with a changing wind, and having to keep a complete log of positions.

The weather remains beautiful and we are praying it will continue. We can finish in 4 or 5 weeks if we have half a chance. However, last December they flew seven days here at Kingsville, so I'm making no plans. . . .

November 1, 1944

Dear Phyl,

. . . Incidentally, I started the month off right in flying, too, getting an "up" on the division check. That's the last official "check ride" for a long time. You just have to get an "up" on each individual flight. Otherwise it was a sad day at the flight line. A couple of the boys in 14C met at 500' last night with disastrous results, and an Ensign cracked up a J this afternoon and got pretty badly banged up. Just three more reminders to take it easy and play safe as possible.

November 6, 1944

Dear Phyl,

. . . The past five days we've done practically no flying. Today was our first since Friday. We had our No. 1 over-land navigation hop, flew 3-plane sections, alternating the lead. With us, as passengers, were three brand new mechanics with about 15 minutes previous flight time. They got quite a kick out of it—and so did we. The flight was satisfactory (we didn't get lost) and my sector came out right on the money. We may fly our five remaining navigation hops now, or we may go directly into gunnery. . . .

Had another Nav hop today, quite successful, and night flying tonight. The latter I don't like, but hope all goes well.

November 6, 1944

Dear Folks,

I guess a good place to begin would be the box of chewies which has just met a sad, but thoroughly warranted end—in the collective gullets of six hungry cadets. Both chewies and mints were delicious and greatly appreciated. I was also quite interested in Dad's enclosure about the Church building fund and the announcement of H—— L——'s wedding. I don't know if I told you, but he washed out in Primary at Olathe just before I got there. Lucky fellow to get another chance at a commission—they're not easy things to get these days.

Texas weather hasn't been all it should be, this past week. We've had fog in the mornings and lots of low cumulus clouds in the afternoons, greatly curtailing flying. However, this afternoon we managed to squeeze in our first navigation hop—over-land.

We flew in 3-plane sections, each man leading the section in a sector search from an imaginary ship, so that our course looked something like this:

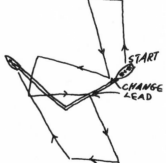

I had the first search and was quite pleased when we came back at the right time and at the right place. After 17 months of practicing this kind of thing, it's nice to see it actually work out in the air. However, flying formation and navigating on a plotting board on your lap isn't the most convenient or easiest thing. To make things more interesting, we each had a mech along as passenger. Mine had been up 15 minutes previously and got a big kick out of the hop and made it a lot more interesting for me. We couldn't do much flat-hatting, but we showed them some good tight formation flying on the way back.

We had no instructor or chase pilot, so it was all ours and we flew it the way we wanted to. We have two more over-land nav. hops and three over-water flights.

We still have the formality of one more division flight to get thru, but should get into gunnery sometime this week. This advanced division work in 2-plane sections is great sport—for example: we fly like this: when going straight or making slight turns. Then when we start on sharp turns we go into what is known as ABC, like this: turning and crossing under so that after the completion of 90° of turn we look like this: each man staying on the inside of the turn long enough to keep up with the formation without using excess throttle.

I was in at the Main Base Saturday and saw several old friends—one an old 6th Batt man is graduating tomorrow! Flying P-boats, though, so I don't envy him. I also saw several of my Corpus roommates who are in P-boats, and they definitely don't like the old wagons. . . .

November 8, 1944

Dear Phyl,

The election being over, things are quieter around the barracks, if not happier in all quarters. Sixteen years with Franklin D is rather a long time.

Last night's flying passed uneventfully. I didn't enjoy it but had no trouble. None of my landings were exactly good, but all were safe for night landings. Today we had our last over-land navigation. I tracked one man out for 65 miles and was supposed to return to a rendezvous point. At our arrival time we were directly over the biggest sand dune in Texas. About that time we see two of the other boys beating out a course at right angles. Contacted them as they went by, and they said they had seven minutes more to go. So I resumed my original course and soon saw two planes circling a town. I thought we were in like Flynn, but soon found these guys, the last pair in our flight, were as lost as we were, so I took a course in the general direction of Kingsville. To make it short, we got home O.K. and found there had been a wind shift of 90° and 20 knots. Our instructor just laughed and said, "Well, you can get away with that now, but wait 'til you start over-water nav!" Nice guy—pleasant thought.

November 9, 1944

Dear Phyl,

. . . We had a full day of flying, starting with a Link navigation hop in the morning. One more final in Nav., three over-water flights, and I'm supposed to be a pretty good navigator—I hope.

This afternoon we had three gunnery hops. It's really a lot of fun and takes a lot of concentration, hard work, and practice. Our first few gunnery hops are "dummy runs," that is without firing. From the high-side run, you get into a 250 m.p.h. dive, and that . . . is really moving.

It turned cold last night, and today was perfect for flying. These three hops put us back on schedule.

November 11, 1944

Dear Folks,

It's a quiet overcast morning with few planes in the air and few people active around the base—an ideal time to catch up on sleep and correspondence.

The last two days have been busy ones with three flights each day— a pleasant change after a week of poor flying weather. These six flights included our last division hop and five gunnery hops. The division hop was awful—we were all tired and out of practice and didn't much care how it looked. I was leading the final section and got cut out and left on the take-off and had to pour on the coal to catch up with the rest of the formation, which added to the confusion.

Gunnery is a lot of fun and also a lot of work. All of our flights so far have been "dummy" runs—that is without ammunition. We are making high-side runs from 1200' above and ahead of the sleeve, coming down in a firing dive at about 250 m.p.h. It's lots of fun but takes some careful planning and good flying. Tomorrow we should have our first firing run. All firing is done over water, so we'll have to lug along overwater seat packs and "Mae West" life preservers. It doesn't seem possible, but we have only about 35 more flights to get in here at Advanced. Friday we had our last Link navigation hop and so, with the final this coming week, will be thru with all but actual flying navigation. . . .

The old gang who started at Liberty, Mo., are beginning to see the end. Paul B——, my roommate at WTS from Chanute, and P—— from Wichita have both finished here and are ready to go on to Pre-Operational next week. Which proves it can be done. . . .


November 14, 1944


Dear Phyl,
. . . We had a lot of fun today on a strafing hop. We go out to Padre Island, the long strip of land that borders the coast from Galveston to Brownsville, hedgehop at about 20' and fire a few rounds at some mock airplanes. It's really a big thrill to tear along the beach at 180 m.p.h. and then pull up to 1,000 feet in a steep wing-over. We're due for some more gunnery hops mañana.


November 15, 1944


Dear Phyl,
Another quiet day with low ceilings, which means little flying—one instrument hop to be explicit. However, we are rapidly approaching the end of ground training. Today we were thrown a test on the mechanics of the SNJ that was really rough, especially as we didn't know it was coming and had absolutely no preparation. I managed to get through it, but that was all. However, about 1725, the temperature dropped 15°–20°, which means the approaching front has passed, already the clouds are leaving, and tomorrow will be clear and bright.

November 16, 1944

Dear Folks,

This letter will contain a minimum of news about flying, for we have had a minimum of it. The entire week has featured low ceilings accompanying a terrific cold front that has made it most uncomfortably cool wearing summer uniforms. Today the Batt office finally broke over and let us wear our leather flight jackets, but the pants are still thin. The new group of Batt officers are doing a pretty good job of messing up our un-military life here of late with inspections, parades, musters, etc., and then they won't let us dress for the weather. So we gripe. . . .

If we get paid tomorrow, and it's a big if, three of us are going down to Brownsville on liberty Friday night and over to Matamoros Saturday. Most of the cadets here get down at least once, and it will be something different at any rate. Biggest purchase for cadets across the border is unrationed shoes—specifically high, brown aviation boots. Don't know that I'll buy anything, but I might have a lot of fun shopping. May also try a smattering of high-school Spanish, like "cuantos" (how much?) and "no dinero" (No money).

Our date of rank came thru today for the 16th of December, which means that if we finish before then, we will be a cinch to go to Pre-Operational. This I don't care for, but there's not much that can be done about it. . . .

November 16, 1944

Dear Phyl,

This regular evening visit with you is one of the high points of the day, or to be accurate, the high point of the day. Especially days like this when we sit around all day and accomplish next to nothing. The bright morning we anticipated, failed to materialize, and it was colder and cloudier than ever. So we played bridge from 0800 to 1230.

This afternoon we had a big review and stood at attention in summer work uniform and winter flight jackets while our hands grew stiff with the cold. . . .

November 19, 1944

Dear Phyl,
Another grey day to keep us on the ground. However the overcast is breaking up slightly and signs are good for our flying tomorrow. Everyone is writing a letter this afternoon as a sort of ritual before night flying. Of course nothing will happen, but we always do it anyway.

Friday night turned out cold and rainy, so our proposed trip to Mexico was scrapped and we went in to Corpus, combining a little pleasure with the business of uniform purchases.

November 20, 1944

Dear Phyl,
. . . We finally saw clear skies today and got in a pair of gunnery hops, which brings us down below the 30-more-flights line, and we should have good weather for some time now—I hope. Incidentally, we didn't fly last night and so still have that to undergo some time this week.

November 22, 1944

Dear Folks,
Well, I have often said that sometime in this program I wanted to find a place where they fly a lot. Today I can say that I've had enough. Yesterday and today we have flown nine hours of really strenuous stuff— three hours of gunnery opposite-course runs, two hours of night formation flying, and four hours of dive bombing.

Last night we had our first of two nights of three-plane tactics. Night flying is bad enough by itself, but with the addition of flying wing on an instructor and flying until your gas tanks are practically empty, makes it sheer torture. We flew from 9:30 to 12:30 and got to sleep in this morning until 0900. Then we had our first four bombing hops, which are pretty rough.

In dive bombing here, we start our runs at 2500 feet and go down to 1000'. Not much of a dive, and it doesn't give much time to correct for aim. We have to get 7 out of 25 bombs in a 100' circle, and it has most of us sweating. We made 10 drops today and got from 5 to 1 hits. I fell in the middle with 3 hits. If I can keep up this average I'll be all right, and I'm hoping to be able to improve it a little. It's the toughest thing we've had to do yet, and there's a large factor of luck in it as far as I can see. At least with all this flying, we are whittling away at the number of remaining flights. . . .

You asked about gunnery firing hops—on the SNJ's we fly, there is one fixed gun (.30 cal) mounted on the engine cowling. We fire 80 rounds on each hop. Firing is done by the pilot with the trigger located on the stick, sort of a pistol-grip affair.

. . . Speaking of bonds, reminds me of the big drive the Navy had last week for extra "Dec. 7th" bonds. Each cadet was asked, and expected, I guess, to buy a $25 bond. The result was that about 5% or less of the cadets did. Most of us said, "Tough, but on our pay, we're lucky to be able to buy one a month, not to mention buy uniforms in the near future." Needless to say, that was my sentiment, too. . . .

November 22, 1944

Dear Phyl,

Boy, oh boy, do I feel lousy! If this letter looks a little more lopsided than usual, blame it on last night's activities. Between 2200 and 0130 I did the hardest, most tiring stretch of flying it's been my duty to do. We had our first night section hop. For two hours we fly around doing echelons on the darkest night in November. Besides trying to see the lead plane all the time, we had to sweat out our gas supply down to the last drop. That, on top of a regular full day of flying, makes me feel like I'd been out on one heck of a big drunk. For once, no one in the flight is anxious to fly this afternoon. Best part of the night was that 14A didn't kill anybody—a couple of ground loops, but that was all. Squadron 14C has killed a cadet the last couple of times they've had night flying, and we all figured it was just about our turn. Today we start firing runs in gunnery, and have about 25 more hops. Had an instrument hop yesterday and had a good look at combat tactics as my instructor and another spent the hour dog-fighting and doing slow-rolls, loops, Immelmans in column. Great sports, these instructors, but it's a poor way to learn instrument flying.

November 25, 1944

Dear Folks,

... Yesterday the weather went bad on us, or rather got worse, and we did one instrument, and five glide bombing hops. This last has us all sweating a bit. I believe I wrote you that we have to get 7 out of 25 hits. So far we've had 15 drops and I have 4 hits. Should have no trouble making it, but there's a lot of luck in it. One of our boys has only 2 hits, another 3, so I'm not too badly off. It would be great fun if you didn't have to get hits and could practice and experiment, but as usual, the heat is on and it's hard work—but still fun. We have a total of about 25 more rides—or less. Sounds pretty good to keep whittling them off.

About this "date of rank"—that means that one may be commissioned on that date or 15 days on either side of it, but actually means nothing. If you don't finish in that time, you wait until a new date of rank can be forwarded from Washington—just a lot of Navy red tape. Pre-Operational is just two more weeks of cadet training, only flying SBD's for the experience of heavier planes. The Dauntless weighs about four times as much as a J, and has 1250 H.P.

Thursday we had Thanksgiving dinner. The Navy did itself proud and, for once, fulfilled our expectations—cream of tomato soup, turkey, ham, dressing, candied yams, peas, green salad, olives, pickles, celery, nuts, hot rolls, coffee, pumpkin pie, fruit cake, ice cream, cranberry jelly, a box of chocolates, apples, oranges—well, that gives you a rough idea of how we fared. Hope you did as well on quantity, but I'd have given a lot to have had the same dinner home-cooked. The turkey and ham were rare treats—we never have ham and our fowl, labelled "spun-in sea-gull" isn't like anything I ever saw before.

We are getting our ground school finals out of the way. I got a 3.8 in gunnery and have checked out of code-sending and flag hoists. . . .

November 26, 1944

Dear Phyl,

The days pass slowly, but we keep whittling away at the flight syllabus. Tonight we have 21 more hops and with good weather might finish by the end of next week. We had our recognition final today, which leaves only two more finals, and then it's all flying. We had one dive bombing hop today, and it was awful. The entire flight dropped 20 bombs, and there were only 2 hits; I'm thankful to say one of them was mine. Now I need 2 for 5 to qualify. The last ones will be all downwind, so I'm not too worried about it.

We are all very much disgusted about a new ruling outlawing Navy approaches, which are short, close, and hot. Because our two squadrons, 14A and 14C have the worst accident record in the entire NATB [Naval Air Training Base], we now have to make a standard civilian approach, wide and square with a 1000' straight-away. The idea was to eliminate ground-loops, but as our instructor, who dislikes it as much as we do, put it, there will probably be twice as many cadets spinning in at the end of the runway. Pleasant thought, and we hope there'll be a change of policy before our worst fears are realized.

November 28, 1944

Dear Folks,

Not much news, but the box of hermits and mints arrived today, so I want to get off a prompt thanks, and boy are they good. We are eating them with Delicious apples this evening.

Clouds and rain today, so no flying. Instead we had a full day of ground school. Now we have but one more thing—a navigation final, and we'll be all through with the "studies."

We only have 19 more flights left here, but despite the fact that no one has yet qualified in glide bombing, we are laying plans for our Flight Party. This is a standard affair for all flights when they finish. We'll probably have ours in Corpus—big dinner and stuff. . . .

December 1, 1944

Dear Phyl,

. . . Yesterday we were scheduled for a couple of gunnery firing hops. The first one didn't come off because the tow plane got his tow fouled on the tail wheel, and we couldn't fire. He had to land with the tow attached, but did a swell job of it. On the second hop my gun wouldn't work, so it was altogether a rather poor day of flying. One flight less, though, and only 16 more to go.

December 3, 1944

Dear Folks,

Foul weather has besieged us for the past few days, and flying has been at a minimum—one hop in the past three days. This afternoon we got in an over-water navigation flight, over land. We flew in two sections; one flew a long flight, and well, but got a "no good." Our group flew a short hop, poorly, and got "O.K." on it. We were supposed to have had a full day of combat tactics and night flying, but we were lucky to get in the one hop. That leaves us a total of 14 flights to go. . . .

There's a lot of new scuttlebutt out about more lengthening of the program. One version is 6 weeks of ground school; the preferred version is 6 weeks of cross-country, flying from Beeville. If that were to work out, and I wouldn't mind it although I have no love for Beeville, I might be able to fly up to Wichita some time. Needless to say, nothing would make me happier than to pull into Municipal Airport some afternoon on an RON hop (Remain Over Night). . . .

December 9, 1944

Dear Folks,

I realize I haven't written much this week, but much also has been happening. We have been working away at flights and have the number down to six—instrument check, combat check, one nav. hop, and three gunnery hops.

Last Wednesday night we finished night flying with a two hour solo section (3 plane) hop. It was really fun, and we flew a very good formation, when we felt like it, just to prove we could.

We have had two of our over-water nav hops over land, I believe I wrote (or maybe I didn't) about the first. We flew out a little ways into the country, then up and had a look at Corpus and the bay. Really beautiful from the air, and the first time I'd had a look at it. Our navigation was off from the start, but not too badly. Today we had the second one, and it was really good. All three of us hit our points on the time and right on course. I had the middle cross-wind leg and was happy when I made my town, because only two minutes before the leg was up, we passed near a large town, and I was expecting that it was the latitude and longitude I was aiming for. But you've got to trust your navigation, so I kept on for two minutes, and at the end of that time there was my town, so small you couldn't see it five miles off, but the right one. The next and final nav hop will have to be over water. We're scheduled for it tomorrow. When plotting, it's a bit rough to hold a formation, and as the instructors usually gripe about our poor formation while plotting, we gave the boy today a sample of how well we could fly. The standard interval is 50' from cockpit to cockpit; well, we narrowed that down to a bare 25', which means the wing man's prop clears the leader's wing by about 2', and your wing is almost directly back of his. We are proud of our formation flying, if little else, and it tickled this chase pilot so much to see a really tight formation that he came in and flew fourth man with us so we looked like this—

I'm not exaggerating distances—it was *close!*

We have had three of our four combat hops. The first one is practice against an instructor, the next two student against student. We gave our boy a fight for his money, and I succeeded in getting on his tail. He was quite complimentary. Said we flew too close to him, though, and scared him. He should have been close around on the next two hops and he'd have seen some really close calls. We miss head-on collisions by about 15' and think nothing of it, and one of the boys about took my wing with him on a join-up that scared us both— but it's really a test of your ability and aggressiveness, and the most fun I've had yet in flying.

The flights start with the two planes flying on opposite courses at 9,000' and 10,000'. As they pass, the top man makes a turn in either direction and the planes go into a "scissors" looking something like this: When the turns become nearly simultaneous and at the same spot, the man with the altitude advantage will find himself on the other's tail. Then a tight climbing circle is started, the lower man trying to climb above and inside his pursuer. If he does so, he can either work into a firing position or back into the "scissors" with the altitude advantage. This sounds very easy and simple, but is actually as vicious a form of exercise as you can imagine. You fly at full throttle and swing the plane around as sharply as possible, all the time trying to stall out the other fellow and avoid a stall or spin yourself. With 90° banks at that altitude, and trying to gain more, the J goes into high speed stalls easily, and you have to fight the controls every second, often using both hands on the stick trying to pull out of a partial spin and not lose altitude. Progressive stalls and spins are nothing, although they were once called the most dangerous maneuver possible. The "ground" is at 6,000', and the first man to go below that is considered dead. You are really sweating when you're in a tight turn at 6,100', fighting off a stall and yet trying to get inside your opponent's turn. It's by far the most thrilling, most invigorating, and yet tiring sport I've ever found. Then to climb back up to 10,000' and see for 50 or 75 miles as easily as you can see a block on the ground—well, that's something worth working and living for.

But then all is never a bed of roses, and one can never, it seems, be free of worry. Thursday we had our final glide bombing hop after nearly a two week lay-off. Coming right after two combat hops, we were in no shape for it, and the whole flight did terribly. Of the entire flight, only 3 qualified, none with extra bombs. Two of us failed to get the required hits, and one has yet to finish, but probably won't qualify—though we all are hoping he will.

I was one of the two who didn't qualify, and had 5 hits for 25 drops, instead of the necessary 7. The other fellow, V——, from Florida, got only 3 hits. Consequently, today our jackets (records) went before the squadron board, and we were given extra time, with 15 bombs to get the 7 hits—all down-wind drops. This will be rough but should and can be done. We are all at odds with our instructor over the deal. It is uncommon for an instructor not to give one or two hits on close bombs, and there have been cases of gifts of 5 or 6, in order that students can qualify. Also, we are one of the last flights having to qualify, and future flights need only drop the bombs wherever they may fall. Thirdly, other flights are qualifying on the first ten bombs with ease, and our entire group had lots of trouble, which, as we aren't great Dilberts, looks like a weak link somewhere. Consequently, we are forgetting about the flight party for our instructor until we are all through with everything. Then if he wants to bring it up, that's his affair. I was quite amused that the fellows who did qualify felt much worse about it than those who didn't.

We were given a chance to look over our jackets—the log of all hops flown here and in Primary, to see if we wanted to present them as evidence of our ability. I am not ashamed of mine in any way. My old instructor at Olathe gave me some fine write-ups along with his cussings-out. My instrument squadron record was highly laudable, my instructor giving me the highest possible marks in tenacity, military bearing, eagerness, adaptability, etc. And all the radio range hops were filled with "fine, smooth, far above average." Down here at Kingsville there were few comments made, the best being on my formation work. I was especially interested in the write-up of my only other NG here, the solo check. Ens. G——, the instructor, and the youngest one on our flight board, wrote in effect: "Student failed to respond to repeated instruction. All landings were wheels first or ballooned. Showed no ability to control plane on ground. Approaches too sharp, erratic, and dangerous. Unsatisfactory in all ways, not safe for solo." Nice one; so next day, as you know, I went up with the senior instructor, Lt. "Garlic" Marculescu, who has better than 1000 hours of instructing to his credit, without a single accident or infraction of rules. He wrote in effect: "Student responded to instruction extremely well. Far above average. Landings excellent. Showed good judgement in all phases of work. Highly satisfactory." The second check pilot, Ens. Yielding, laid it on equally thick. If I ever get my wings and am down here at Kingsville, I'm going to show those three paragraphs to Ens. G—— and see how many colors of red he turns. But first I've got to get three hits in glide bombing. The board gave both of us 15 more bombs and 7 hits without even interviewing us, and the Asst. Chief Flight Instructor said he didn't want to see us again, so I guess there's nothing to do but go out and get them. In order not to hold up the other fellows, we'll probably finish everything else, at least nav. and gunnery, before we start in on the bombing again.

Friday I had to replace a raincoat someone swiped, so three of us went in to Corpus to attend to various and sundry business. And lo, the town is full of ex-cadets whom I have often gone on liberty with, now sporting brown shoes and gold wings. It's really swell to see the boys in their greens, or wearing blues with covered stripes while awaiting Wednesday graduation, and believe me, they are a happy lot. The CV syllabus boys are sitting around Cuddihy now for two weeks after graduation taking cross-country hops while they wait for orders. Rough life—fly when and where you want to, when and if you want to, and liberty whenever you want it. . . .

Dad asked about flying to Brownsville. That's a goodly little jaunt from here, and there are a number of Army fields down there with heavy traffic, MacAllen and Harlingen, specifically, so we don't fly down there very much. A few instructors take nav hops down along the Rio Grande, but our boy got lost down there once and is apparently afraid to repeat the experience. Also the Navy wouldn't care for a forced landing on an Army field or in Mexico. . . .

December 11, 1944

Dear Folks,

Just a note to keep you posted on current events. Today broke cold and clear, and we were very glad to go into greens at long last. We've been shivering around in khakis much too long.

Three of the fellows went out on their last over-water nav hops but couldn't get contact with Corpus Radio and had to incomplete the flight. V—— and I had our bombing lecture and two hops, the third fellow, H——, who hasn't qualified yet, going on the second one. On the first hop, he got 3 hits and I got in 2. The second hour we figured we had the gouge on it and expected 4 or 5 hits apiece. We did not figure on a gusty and varying wind, though, and that threw them and our instructor for a total of 0 hits. I was lucky and got 3 hits. Our new glide bombing instructor, a Marine lieutenant, incompleted the last hop because of the sparcity of hits, and so we still have 10 bombs left. On the next hop I will get credit for the 3 hits I got today and will have a chance to raise it, but won't get less than 3, so it's a break for me either way. . . .

December 12, 1944

Dear Folks,

Well, I had a big day today, and finally have the glide bombing behind me. We went out on our last hop needing 2 and 1 hits and got them right off the bat. Sure was a good feeling to see those bombs burst right in the bullseye.

Also completed ground training today with a rugged navigation final. Did pretty well, I believe. Three of the guys got their over-water navigation in this morning. One of them went off the runway on his landing, but did no damage to the plane. . . .

December 17, 1944

Dear Folks,

This is a red letter day, as this day I did complete Advanced Squadron. Yesterday a combat check and gunnery hop. That last gunnery hop about threw us for a loss. In the process of getting it in, we lost four tows. Today, despite lowering skies, we flew our instrument check and last over-water nav hop. All went well for everyone, and tomorrow at 0800 we start checking out. It's a mighty good feeling and one we've often doubted we'd ever experience. The box of fudge came thru in good shape; although it didn't last long, it was appreciated by all, especially yours truly.

We'll leave for Main Base on Wednesday or Thursday, lie around there 'til Saturday, then out to Cuddihy for two weeks of Pre-Op in SBD's. From all we hear, it's a pretty good deal out there. We should get our wings about the first week in January. Seems impossible, but I guess it's actually all coming true at long last. These 18 months have gone quickly in ways, but in others very, very slowly.

Phyllis is spending part of her month's Christmas vacation down here, so my liberties are currently improved. One of my roommates also has a girl in Corpus, and we make the trip to and fro together.

At present the newly commissioned ensigns are waiting around for two weeks at Cuddihy for orders, flying V's on cross-country hops, then getting orders with travel time *and* 28 days leave. Sounds pretty good, and flight pay begins immediately upon graduation. That pay check is going to look about as good to me as the wings when it starts carrying three figures on the left side of the decimal point.

The next few weeks will be pretty hectic, but I'll keep you as well posted as possible on where I am and what I'm doing. . . . *And a Very Merry Christmas!*


Cuddihy Field
December 22, 1944

Dear Folks,

Pardon the pencil, but some place in the rush of moving and packing I misplaced my pen and so must use what is at hand.

To begin at the present, I'm relaxing after my first day in Pre-Op. We arrived late this morning and started to work immediately with a long afternoon of lectures on the SBD down at squadron.

Cuddihy is much like I left it six months ago, many of the same officers around, which is a good thing as they are a good bunch, especially the Batt C.O.

The Pre-Op squadron, VN19D8, is located in the same hangar 12A was, my old Basic squadron. The same course rules are in effect, and it's a sort of homecoming for us old boys. They are flying SBD's, SNV's, and SNJ's from this field. Part of 13C, the Beeville instrument squadron, is down here, and apparently Pre-Op will soon move out there and instruments down here. Typical Navy way of doing things. There are almost as many varieties of cadets here as at Main Base. We have boot ensigns, Pre-Op boys, instrument cadets from Beeville, and a contingent sent in from Pensacola for some unknown reason.

We left Kingsville Wednesday and spent two days at Main Base on almost perpetual liberty. We were bunked in the same barracks we were in as indoctrination cadets—and alongside the current crop of boots. Main Base is doing its best to put on a holiday face with Christmas recordings and a number of night displays.

Apparently the present training program is planned to have all CV cadets train at Corpus and all multi-engine and observation training to be given at Pensacola. No one knows exactly what is happening, and they are making cadets choose their branch of training while still in Primary.

Corpus is more prosperous and busy than I've ever seen it, and everywhere I go in town I run into new Ensigns whom I was once rooming with, flew formation with in Primary, or even left behind some place along the line.

My roommates here are the same old bunch from Kingsville, with the exception of H——, who is being held a week and sent to Beeville for Pre-Op because he is getting through ahead of most of his original class.

I'll wait to tell you much about our training here until I've more facts to write, but here are a few items on the SBD: it's a scout-bomber, built by Douglas, that has seen more action than any other Navy plane. Weight 8500 pounds, length 33', wing span 41½ feet, height 12'10", powered by a Wright Cyclone single-row, 9-cylinder air-cooled, radial engine, giving 1200 brake horse power at 2500 RPM. It uses a Hamilton Standard, 3-bladed hydromatic constant speed prop with 10'10" blades. Fewer Dauntless have been lost in action than any other plane, and they are admittedly the best constructed plane, tougher and more rugged than any other in the Army or Navy. Which is, you might say, a lot of plane!

. . . Purchased a very beautiful set of blues yesterday. Got tired of waiting for the Ship's Service store to get my size in stock.

We are having a pretty swell Christmas liberty—1800, 24th to 2200, 25th! My roommate H——, his girl J——, Phyllis and I are planning on a big dinner party at the Driscoll—which will be the next best thing to being at home on Christmas eve.

A very Merry Christmas to you—and a Happy New Year, and my thanks for the best Christmas present possible—excepting that it be at home.

December 26, 1944

Dear Folks,

Well, the big holiday is over, and we are more or less off to work again. Our three days here have been spent listening to lectures on every phase of the SBD. We're really getting the word on how the plane flies. This afternoon we had our pre-flight test—passed O.K., and if the weather permits, we'll fly tomorrow. However we had fog and rain today, so it looks bad for tomorrow. The training here is much better than any we've previously had. The equipment is better, lots of mock-ups of hydraulic systems, controls, cockpits, and all the instructors are flight officers, understand our viewpoint, and really want to give us the word.

The base is much like it was this summer, easy on inspections and military, altogether good living. I'm rooming with four of my Kingsville roommates. The sixth boy got held over a week because he is in an August class and hasn't been around as long as most of us.

The only rub is that the squadron is moving to Beeville next Monday. We are hoping like everything that we can finish here and won't have to go out there again. I don't know what our chances are, but if we go out to Beeville, there's no telling how long this will take as they are stretching the program from 7½ hours to 25 hours of flying and lots more ground training. . . .

This noon the Cadet Regiment kicked in with our Christmas present—a box of candy and a sack of nuts—for the $18 we have paid for past 6 months.

Which seems to take care of events to date. All now is up to the weather—which is no new story. Latest orders for CV graduates were F6F fighters and 28 days leave. Here's hoping!

December 29, 1944

Dear Folks,

I shall write under a slight handicap of not knowing just what's going on. I wrote about the proposed movement of 19 Squadron to Beeville, you'll remember; well, this morning the entire personnel packed for movement, about three-fourths of the cadets left, and the other fourth spent the day going to ground school. I am obviously still here, bags still packed, and expecting the unexpected. No one has told us we will spend the night here, but as it's 1800 now, I presume we will. Today we had the entire Navigation course and two hour final, and review for the Recognition final tomorrow. No flying is being done, due to the weather. We don't know whether we will stay here to finish, fly up to Beeville, or what. It seems fairly certain we'll be here long enough for the Rec exam tomorrow, but beyond that is sheer guesswork. This is a perfect example of the Navy's organized disorganization. On top of which, about a hundred instrument cadets came down from Beeville today. There is much talk about the new syllabus for Pre-Op—25 flights including navigation, formation, and an oxygen hop.

The Christmas packages arrived yesterday and certainly added a touch of home. . . . Besides having two days off at Christmas, the following Wednesday night and Thursday were our regular liberty days, so with Phyl here, it made my Christmas as festive as possible. Thursday night I saw Phyl on her way to Maine again.

. . . I sent out very few cards this Christmas. . . . I'll also feel a lot more like writing when I get this program out of the road and can write that I am no longer "in the last phase" or have "only five more hops" no longer, but write that Ens. Rea is waiting for orders at present.

We are all thoroughly sick and tired of this cadet life and ready for a change for the best. The present situation doesn't help morale any either, so I'll close and catch up on a few of the other letters I owe.

Cuddihy Field
December 29, 1944

Dear Phyl,

. . . This has been a fouled up day. Last night when we got in from Corpus we learned that the Right Wing (the other half of the squadron) and all pool cadets were going to Beeville this morning, and we would go with them if the weather was bad. The weather was and is bad, and we packed everything, only to be called out at 0900 for navigation. In three hours we had the Nav course, condensed slightly, and then took the final this afternoon, along with a review for the Recognition test tomorrow. About 100 cadets came down from Beeville, and we are the only SBD cadets left here. Nobody knows how long we'll be here, what we'll do tomorrow, how much flying we'll have to do, and we are ready with our bags packed for whatever may come. And I mean we are a very P.O.'ed bunch of Cadoodlers.

Chase Field, Beeville, Texas
December 30, 1944

Dear Folks,

Now we can say with McArthur—"I have returned." First Cuddihy, then Beeville; I fully expect to go back to Kingsville before I leave Texas.

This morning we had our Recognition final and came out here at 1300. The final was quite easy. I got a 3.9. Had a 3.7 on the Nav final the day before.

The weather is just plain lousy. Almost constant mist, fog until noon, and then you can't tell whether it's fog or just the bottom of the low stratus clouds. About 1300 we could make out the sun through the clouds, but little sign of a break in weather.

No one knows what we'll do out here or how long we'll be here, so we're just sitting around, taking it easy, and finishing off Christmas cookies and candy as fast as possible. Don't expect to get any mail for a week, and by that time we hope to be back at Main Base. The SBD's are still at Cuddihy, and all the cadets here.

Ran into a couple of my old buddies from Olathe this evening at chow—hadn't seen them since June. Chow here is pretty good—better than Cuddihy. I'm even living in the same barracks I was in here in August and September. . . .

Chase Field, Beeville
December 30, 1944

Dear Phyl,

. . . We came out to this god-forsaken hole this afternoon after completing ground school at Cuddihy with a Recognition final. All went well, and my ground school average for Pre-Op was 3.8, which satisfies me to a certain extent.

The weather remains foul—fog until noon, mist almost constantly, and no sign of a break in the clouds which seem to almost touch the ground at times. No one out here knows a thing about what we'll do here, so we are just sitting around waiting for something to happen.

December 31, 1944

Dear Phyl,

Well, I went to church this morning, and lo and behold, the sun is now shining brightly and there are only a few little clouds in the sky, and they are running for their lives. About noon the instructors began ferrying J's to Cuddihy and SBD's to Beeville, so we now have about half our planes here and the rest on their way. If all goes well, we should have all-day flight tomorrow and finish flying on the 2nd. This, of course, if we don't get the long syllabus.

January 3, 1945

Dear Folks,

The past couple of days have been fraught with fruitlessness in all lines of endeavor, so we will pass over them quickly. We have been spending eight hours a day down at the line waiting to fly, and never flying. Yesterday I got to taxi a plane and considered myself fortunate. . . .

Today I managed to get my name on the board and was given the last spare SBD for my first solo. The flight went smoothly and uneventfully. Outside of attempting to taxi with a locked tail wheel, I was quite pleased with the flight and the plane. It's a cinch to land and take-off and handles nicely in the air. The Dauntless is heavier and slower on the controls than the good old "J," but there's a lot of power there when you want it. Did nothing but fly around experimenting with power settings, stalls with flaps and wheels down, and flipper turns and accidental stall recovery. The next period will be landing practice, then a sort of landing check and one final solo.

P.S. Enclosed ad shows SBD in battle paint. Ours are the old deep

SBD Douglas Dauntless divebomber flown in Pre-operational training
at Corpus Christi (Courtesy U.S. Naval Institute,
Library and Photographic Services)

blue with white bellies and are always badly grease and oil streaked
from stem to stern.

January 4, 1945

Dear Phyl,

It is now 0649½, and so I am still in somewhat of a daze, and it's
too early to get P.O.'ed yet, so I'll start a letter to you and take my
chances on finishing it later today.

The dirty dope is that we were given a new shaft even at this late
date. A group of the Waldron boys led by our chief junior bird-dog (cadet
officer) got the ear of Schedules and fed them some line of corruption
about having to graduate by a certain date, and got their names on the

schedule board for two straight days, excluding everyone else. Consequently, these 33 men are going in to Main Base today, and the rest of us are stuck out here for another week. This wouldn't be so bad except for the fact that 28 of these 33 men are in classes 7A–7D, one to four weeks behind us, and have missed the pools that held the rest of us back so much. To say the least, we have a bunch of disgruntled, disagreeable cadets on the base.

One of our main difficulties in getting hops in is the condition of half the planes in the squadron. We have about 100 SBD's, and there are never more than 50 in the air at one time. The rest simply can't be kept in shape to fly. We are averaging one good crack-up per day, and a couple of minor ones to boot. Tuesday two instructors and four cadets got lost and landed up to 175 miles away.

By waiting around the schedule board for an hour and forty-five minutes yesterday, I managed to pick up the last spare plane and get in my first solo hop. It took half a dozen mechs, a j.g., and myself to get the old crate started, but I got it off the ground O.K., and had a pretty good flight. The Dauntless handles easily and smoothly, if heavily, and has no unpleasant characteristics outside of a vicious steep turn stall to the left, from which you have to recover by rolling, and lose 700'. Actually, it flies better at 70 knots than the "J" did, and lands like a dream.

This sitting around waiting for flights is pretty hard on my nerves, and we are playing bridge or solitaire like maniacs—even I am now working crossword puzzles upon occasion. We'll be here until the 11 Jan., and will graduate 17 January if nothing more comes up to delay us. . . .

My chances for instructor's duty are practically nil now. The list of recommendations came out yesterday. Barstad named McG—— in our flight, and he, after much indecision, decided not to take it. Not very brainy for a guy who wants to get married, and it screws the rest of us out of a crack at it. Three of us would have jumped at the opportunity. But that's life. . . .

January 7, 1945

Dear Phyl,

. . . Squadron is as fouled up as I am. First we have bad weather, then liberty, then I fly my second solo, then they start flying an entire new bunch, and tomorrow we go off of all-day flight and only have half as much chance to fly. Scuttlebutt is so thick you have to keep your eyes closed in order to see during the day, and I think new dope gets started at night when the guys talk in their sleep. F'rinstance—the latest word is that 92 of us will go in to Main Base Thursday, regardless of flight time, and/or we will be here at Beeville until we have had 15 hours, which at the present rate will take 7½ weeks. Nobody knows nothing and everybody is talking, so I've given up all plans, hopes, or ideas, and wouldn't be surprised if we graduated the 17th—either January or March.

So much (which I'll readily agree is too much) for this. . . . You will be happy to hear that I purchased for you two (2) rolls of film which may or may not fit your camera. They are supposed to, but if not, you should be able to extort gold for them in Portland.

January 9, 1945

Dear Phyl,

Recent changes have bettered our flying lot a bit the past couple of days. Yesterday morning we had two hops—a landing check, which was a farce, and a solo. I went out with P—— on the last hop, and we flew a two-plane section up to the Army air field at Victoria, Tex. Not wishing to tangle with P-40s flying in that area, we peeled off and went down on the deck, dodging snags in the swamps around there. After chasing a few cows, I pulled up in a wingover, but P—— stayed down so I went after him and we headed down a little valley in column. Great fun flying below the ground level until you hit a little slipstream. Then we pulled up to about 3000' and located a busy highway intersection and dive-bombed passing traffic. After a couple of runs we headed back north across the prairie flying wingtip to wingtip at what would be called a low altitude. Some people consider anything below 500' low, but when a couple of safe and cautious pilots start flat-hatting, they do a good job. Had we lowered our landing gear, we could have rolled our wheels on the ground. We were pulling up for fences, not to mention trees and telephone wires. After about half an hour of that we pulled up to 100' and headed home. About that time we had a good scare when we spotted another SBD tailing us. We figured at once that it was an instructor and our goose was well-cooked. But our luck was good and it was another cadet who joined up with us for the return trip home. It was by far the most fun flying I've ever had, and there is nothing prettier than a couple of planes skimming along at 240 knots, wing to wing, and nothing more thrilling. Today we are scheduled for a section tactics hop. We will at least get started on the new syllabus, whether we have to stay here another week to finish it, or not. I'll be glad to leave as soon as possible, not just to get the bird, but because this battalion is rapidly becoming strict as they can be on matters of room inspection, military, etc. Yesterday the inspecting officer put over 100 men on report, for which there is no excuse or reason. Everyone is in a foul mood, and never a day passes that we don't get some new shaft thrust in us. Oh, yes, you asked about the meaning of "P.O.'ed." That's an abbreviation for a rather crude, but very descriptive Naval term meaning about the same as "disgruntled."

The foregoing was written this morning early. Our scheduled flight failed to materialize due to a shortage in serviceable planes. Did little this afternoon but listen to some lectures and see a movie we had already slept thru once.

Main Base, Corpus Christi
January 11, 1945

Dear Phyl,
    ... The situation at present is this. At 2400 this evening our date of rank expires so we are stuck here until a new date of rank comes out of Washington. Thus we will not graduate on the 17th, and will be lucky to get out the 24th. In the meantime, we just sit around and wait. . . .

I am so damned tired of the whole thing this evening I can't write you a decent letter, Phyl. It just isn't in me to be happy or cheerful. Some of the guys are mighty sore about this latest shaft, but I can't get very excited about it, merely more and more disgusted. We were given an instructor preference card this afternoon which may or may not mean anything. I said I would accept Basic or Primary instructor's duty, but doubt, with this latest shaft, if it will have any effect. Just so I don't get dive bombers—I've flown SBD's as much as I want to.

Main Base
January 13, 1945

Dear Folks,

Time passes, but the Navy never changes, and we are still getting lousy breaks with steady and unvarying regularity. The latest just about takes the cake. We left Beeville Thursday after spending 3 weeks in Pre-Op and flying only 7.4 hours. Thursday evening we learned that due to the innumerable delays that plagued us since leaving Kingsville, we had missed our date of rank and would not graduate the 17th, but maybe the 24th. I don't need to elaborate on how disgusted we were. But there's nothing that can be done about it, so here I sit, without a thing to do for a week. If I could have gotten a ride north, I'd have taken a week's leave and come home, but there was no available space on the NATS [Naval Air Transport Service] plane to Olathe, and I couldn't see the train ride and expense when there's bound to be leave with our orders, and much more, I hope.

Yesterday we had our physical fitness check-out, the only thing we were able to check out in this week. Today I've more sore muscles than I ever dreamed of. I did pretty well on it—among other things 12 chins and 31 push-ups, which is better than I did at Preflight when we were supposed to be in our best physical condition. My weight is at an even 130, so I guess I'm in pretty fair shape.

I don't believe I wrote you the details of our last flight in SBD's. It was an hour and a half solo, so one of my roommates and I flew a two-plane section northeast of Beeville to the big Army field at Victoria, Tex. They've got a lot of P-40's up there which we didn't want to tangle with, so we peeled off at 2000' and went south toward the Gulf at ground level, dodging trees in the swamps, going down ravines and gullies, until we hit Matamoros Bay. Then we pulled up to 3500', dive-bombed a busy road intersection and headed off across the prairie flying wingtip to wingtip, and low enough that our wheels, if lowered, would have been rolling on the ground. We kept well over 200 knots for about 15 minutes and covered ground fast. It was really beautiful to see your wingman right there beside you, pulling up for trees and fences. The SBD is the ideal plane for hedge-hopping, as it's heavy enough that gusts and rough air near the ground don't rock it or cause it to bounce around. One of the fellows went down to Kingsville and took on a J in combat—did O.K. until another J came in to help, then had to run for it. The SBD is about as maneuverable as a 2-ton truck, but when you get it going, it really makes tracks. One of my roommates tried a non-flap dive—got 415 m.p.h. out of it. Best I could do was 365 m.p.h.—they are red-lined at 375. One of the daredevils in the squadron took one up to 21,000' feet without oxygen and diving down got well over 500 m.p.h. The trouble is that at such speed these old planes start falling apart pretty easily, and the only way you can pull out of a dive like that is by using tabs—the controls are practically frozen above 400 m.p.h. It was undoubtedly good experience flying the heavier planes, but I'll be happy if I don't get SBD's for operational.

Every place you go, nowadays, you run into fellows you've gone thru the program with who have their wings. When we graduate, it will be with guys who came down here in August and September. Yesterday I met a fellow who was in my platoon at Liberty, came down here six weeks after I did, will graduate when I do, and is griping because he is having to wait an extra week or so to be commissioned.

Went into Corpus yesterday afternoon and tried on my uniforms. They look mighty good and I am eager to get them on, to say the least. There's still a 10–14 day delay for orders after graduation, so I won't be leaving here until some time in February. I've given up writing to people because there's nothing to say, and most of them have already congratulated me on my supposed graduation. After so long, I imagine it seems pretty funny, but most of us here now have reached the point where we don't much care, one way or the other.

I ran into a fellow from Friends Thursday night. He was an Alpha Kap, played basketball, and got into the program a couple of months after I did. He was in on the big 50% cut in Preflight, made it thru Primary when the going was toughest, got down here, and can't pass the eye examination. They haven't washed him yet, but probably will before long.

. . . The package mailed to you from Beeville is a little of the superfluous gear I'm still carrying with me and not needing at present. I'd appreciate your putting it some place where you can send it to me if and when I need it again. I'll probably send home another box with my whites and khakis in the next couple of weeks. After we are issued winter flight gear, plus two more uniforms that I want to keep in fairly decent shape, my bags will be plenty crowded. We will be issued at least one more bag, however, when we graduate.

Sure hope I can see Bill Roy when I'm home. Had a letter from him last week. He checked out in an SNJ when last in Frisco and thought it was a pretty swell plane. He didn't fly J's in Advanced, just PBY's, and so a light, fast, maneuverable single-engine plane was a real treat and new experience for him. These "truck drivers" have the best duty in the world, but I haven't met one yet who wouldn't like to fly the jobs we get to—whether they're boot ensigns fresh out of Beechcrafts, or veterans like Bill who have thousands of hours in the big stuff.

Some Army pilot in a P-47 "Thunderbolt" had an emergency landing at Beeville Wednesday, and we got a good look over the outside of one of the Army's best fighters. Has a 2000 plus HP engine, and practically no wings or fuselage. The engine alone dwarfed the SBD's, and they aren't a small plane. . . .

Corpus Christi, Main Base
January 13, 1945

Dear Phyl,

This morning it's me who needs the rub-down. I mean I've got more sore muscles than I ever had at one time in my life. Reason why— we had our physical fitness checkout yesterday—chins, push-ups, jump reach, obstacle course, step test. I am, or rather I was yesterday, physically fit. Today I'm one sore cadet.

The edge is beginning to wear off our feelings about missing our date of rank, and now most of us are just tired and very, very uneager. If one didn't keep running into guys who have just gotten their wings who were once with you or even behind you, it wouldn't be so bad. The leave deal didn't work out, so we're stuck around here with absolutely nothing to do until next Friday.

Yesterday I went into Corpus and tried on my blues. They are very satisfactory and I am quite pleased.

January 15, 1945

Dear Phyl,

. . . After lunch we went over to the administration building to see if our new date of rank would be in time for graduation next week, and getting an affirmative answer, did what we could to assure our getting the kind of orders we want. With luck, I at least am assured of fighters and will get a chance at instructor if they have any demand for them next week.

Then we saw a lousy show and bought some uniform stuff—cap devices, ensign bars, etc. So far I've spent $17, owe $99, and have $178 coming before graduation. We were figuring up the other day, that the moving of Pre-Op from Cuddihy to Beeville cost us about $150 cash and considerable delay in promotion, if and when. . . . I picked up my shoe coupons today, and I guess I'll have to wear the shoes in our family, Phyl, because I can get a dozen pair of shoes a year, but none of the coupons are transferable.

January 22, 1945

Dear Phyl,

. . . Haven't actually gotten my orders yet, but know pretty well what they are. I am going to Operational at Green Cove Springs, near Jacksonville, Fla., for training in VF, that's fighters. I'm not sure just what we'll fly, but as far as anyone can learn, there's a good chance for land-based Corsairs—F4U's. This would be ideal—good orders, good plane, and good duty. Never know, though, 'til you get there, what it will be like.

Wichita, Kansas<br>January 27, 1945

Dear Phyl,

The big day has come and gone, and this snowy Saturday afternoon finds me at home, finding it hard to believe it's all true. Wednesday, graduation day, was one for the books. We hit the deck at reveille for a change, put on as close to a cadet uniform as we could muster, having previously packed all excess gear. Chow finished, we checked in our linen, shaved, and put on blues, get our picture taken, then while it threatens to rain, we go over to the Rec. Mess and stand at parade rest for 25 minutes until the Admiral arrives. The ceremony was quite

Ensign Robert R. Rea, January 1945

unimpressive—the only big thrill being at the conclusion when Col. Sapp, commandant of the cadet regiment, winds it all up with "Officers, dismissed." At that we let out a yell you should have heard in Portland. We pound each other on the back with suitable comments and much use of the term "Ensign." Then out the door and part with a buck for the first salute. C—— H—— got mine—he was the only cadet I thought deserved a break . . . they kept him at Beeville 4 weeks, so he won't graduate until the 31st.

Back to the barracks, out of blues and into greens, tear over to disbursing and dispensary for health and pay records, then finish packing. P—— and I were sweating out the 2:30 train north, and we just barely made it. . . . The trip was nothing to write home about; I just caught my next train in Ft. Worth and got quite expert at running through crowded train stations with three large bags in tow. Worse than the step test.

So, Thursday night I pulled into Wichita and have been catching up with myself ever since. My orders, as I may have written, are for F4U "Corsairs" at Green Cove Springs, Florida, with 34 days leave. . . .

# 7

# Corsairs and Carriers

**W**HILE NEWLY COMMISSIONED Ensign Robert R. Rea underwent the first phase of Operational training, the war in Europe ground inexorably to a close. Once the German counteroffensive in the Battle of the Bulge was smashed, the Anglo-French-American tide lapped over the Rhine and the Russian masses poured into eastern Germany. In the ensuing weeks, German cities were pounded to rubble from the air. One episode, the torching of Dresden, would lend some credence to the thought that the Western Allies had become as deliberately brutal as the Nazis or the Russians, whose troops engaged in orgies of rape in the towns and countryside of Germany. The Germans' own indelible contribution to twentieth-century savagery was now harshly exposed as the Allies liberated the Death Camps.[1]

Matching the devastation of German cities, B-29s flying from bases on Saipan, Tinian, and Guam wreaked death and destruction upon the cities of Japan. Flying by day and night in a one-sided offensive, they released not only conventional high-explosive bombs but thousands of incendiaries. In one night alone, the death and destruction in Tokyo surpassed that of both Dresden and Hamburg. The rising violence in the Pacific was underscored by the two bloodiest amphibious landings and island pacification operations of the war—Iwo Jima and Okinawa—as the Americans drove nearer to the Home Islands. The kamikazes were especially active against the supporting fleet off Okinawa. There, too, U.S. Navy planes played a leading part in frustrating the last major sortie of the Imperial Fleet. In the process they blew up the world's largest battleship, in a naval vindication of Army air power prophet Billy Mitchell.[2] But the Japanese were far from surrendering, so Bob Rea's training in Florida was taken most seriously.

After visiting with his parents in Wichita, Ensign Rea went to Portland, Maine, where he and Phyl were married on Valentine's Day, 1945. From head-high snowdrifts in Maine to early spring in Florida was as much of a change as that from cadet to ensign—or from bachelor to married man. Before the postwar Navy mothballed a fleet at Green Cove Springs, the St. Johns was a beautiful river, and the town slumbered happily in the early twentieth century. Lee Field was a pleasant place, dedicated to the business of pilot training, and save for separation from a new wife, the life of an officer-pilot was sweet.

The F4U Corsair was an awe-inspiring monster that folded its wings at night. Its mighty engine—thirteen feet of steel in front of the pilot—sprang to life (usually) on the firing of a blank shotgun shell, for these were old models by the standards of the day. And how it flew! How it landed was a matter of some dispute between Navy and Marine pilots. The latter (including the esteemed Captain Leslie) distrusted it at low speeds and preferred "wheels on" landings; the Navy boys took pride in rolling the tail-wheel first. Historically, the "Gyrenes" had some claim to the F4U, but at Green Cove Springs they flew F6Fs, the Navy's Hellcat—and ground their teeth in envy as they watched young ensigns show off.

The flight syllabus was practical, concentrating on formation flying, combat tactics, and gunnery practice. It was difficult to imagine that the fluttering white towsleeve was an enemy plane, but the bullets were real and were color-coded to indicate each pilot's hits—assuming his guns would fire, which was a question usually answered by pressing every firing mechanism (for guns, bombs, and rockets) in the cockpit. It is doubtful, however, that the Navy persuaded anyone that 90° deflection shooting was the best way to down an enemy aircraft.

Combat tactics consisted of the famous "Thach weave." The theory looked good, but the distinguished and decorated instructors did not say much about how it worked in practice. One got the impression that wingmen were expendable. It could also be dangerous, as Rea discovered one afternoon when he had a minor midair collision which went into his flight jacket as "Pilot error—other pilot." Both planes were landed safely, but the wives in Green Cove had heard the word before the pilots returned to town, which led to some fast talking and careful explaining.

Flying became very serious when it came to Field Carrier Landing

Practice. At that stage one learned what it meant to fly "low and slow." In fact, it required one to learn that a Corsair would fly at sixty-four knots in spite of the instruction books, and anything faster was likely to produce a wave-off. There was obviously little room for error.

Typically, Rea's orders to join "Fuzzy," the escort carrier *Guadalcanal*, came without warning, and Phyl was left (not for the last time) to wonder what the Navy had done with her husband. Ensign Rea and his flight were equally in the dark as to where they were, for the whole operation was kept marvellously "hush-hush." Carrier takeoff in the powerful Corsair was routine. Carrier landings were a test of one's faith in his own skill—and the Landing Officer's good judgment. The "paddle" men aboard the *Guadalcanal* were good: calm, efficient, dependable. Daylight Carrier Qualification was completed with only a minimum number of wave-offs, and Rea spent the last day or so aboard an accompanying destroyer escort observing how the saltwater Navy lived.

# Operational Training,
# Lee Field, U.S. Naval Auxiliary Air Station,
# Green Cove Springs, Florida,
# February 27–June 1, 1945

*On February 24, after a honeymoon in Boston, Ensign and Mrs. Robert Rea went to New York to visit briefly with Phyl's brother, Paul Edwards. The narrative in the following letter begins on Sunday, February 25, 1945.*

Green Cove Springs, Florida
February 27, 1945

Dear Folks,
    Long time no write, but I hope you'll forgive and understand. . . . Sunday night we parted at Grand Central Station, Phyl for Portland and I for Florida.

Phyl and Bob Rea

There is but one non-reservation train for Jacksonville a day from
N.Y., but I managed a seat, rode the lounge car Sunday and didn't have
too tiring a trip. Passed thru Philadelphia and Washington at night,
however, and so missed them. Can't say anything good for Virginia, N.
and S. Carolina, and Georgia. They have some of the worst country I've
ever seen. Even made Texas look good. Spent the night at Jax (hereafter
to mean Jacksonville), washed up, rested, and came down with 4 other
C.C. [Corpus Christi] boys at 10:00 a.m.

Green Cove Springs is a very small, quiet, and pretty town, set on
the banks of the wide St. Johns River among thick moss-draped pine
woods. The base is a well-kept, small place also on the river. Our quarters
are quite nice, if plain. Two men to a room—I am with R. S. P——, the
K.C. boy I rode up with from Corpus. We have ample room, bureaus,
closets, and good single beds. The Junior Officers' Mess is quite nice,
but plain.

Our flight got Corsairs—F4U's, and so we are quite pleased with the prospects here. We will be here at least 2½ months and possibly 4, depending on the usual factors. A number of my old friends are here, and we have had quite a time comparing notes on our leaves.

The weather here is amazing. It's beautiful and actually hot enough to wear khaki or greys. And we do have mosquitoes, though not badly at present. Really, G.C.S. seems an ideal place to be stationed as far as surroundings are concerned. . . .

February 28, 1945

Dear Phyl,

. . . A quiet day spent writing letters and completing our check-in. I was quite relieved to pass my physical without any trouble. My left eye is very good, but the right one is pretty weak for 20/20 vision—passed, though.

The liberty here is better than at any other operational base I've heard of, so when you come down, we should have considerable time together. . . . I will have one day, one night off each six days, and only two days a week I can't leave the base. I repeat—come prepared for warm weather. . . .

Just got the word that we are getting paid Friday, and my first pay should hit close to $250, so I'm hoping to get some money to you before too long. My subsistence here runs $30 per month, which isn't at all bad. No MOQ [Married Officers Quarters], so we are also drawing $63 per month rental. And did you know that every day we were married after 14 Feb., the Navy was paying you 50¢ a day to keep their good Naval Aviator happy?—well anyway that's your share of our subsistence check. My travel allotment from CC to GCS [Green Cove Springs] runs to about $86, so financially we should get on our feet again in short order.

March 1, 1945

Dear Folks,

. . . Today we began our pre-flight instruction, and it is quite a grind. We have 4 days of intensive lectures and hundreds of pages of technical orders and bulletins to read and digest before we begin flying on Tuesday.

Yesterday we completed the lengthy process of checking in. My big worry was the physical, but I passed without much trouble. My left eye is in perfect shape, but the right one is just a bit weak where perfect vision is required. So far, so good, though, and we're off on another phase of training.

There are a number of little things I would appreciate your doing for me. I need here the khaki pants I left in Wichita and also the khaki swim trunks in a drawer of my bureau, and also my big hunting knife. This latter not to kill alligators but to puncture seat-packed rafts which have a bad habit of inflating in flight and jamming controls. . . .

The food here is amazing, considering it is furnished for 70¢ a day—the service wonderful. Breakfast is especially choice with selection of eggs any way you want them and no limit. Lots of milk and butter. . . .

*c.* March 1, 1945

Dear Phyl,

. . . I am now well settled at Student Officers' Quarters at GCS. . . . Our quarters are fairly nice. Two men to a room, plenty of dresser and closet space. I am rooming with R. S. P——, the K.C. Kid. Our mess is nice, satisfying, but nothing elaborate, which can hardly be expected for $30 a month. The base is about the size of Kingsville but quite attractive and well kept. I am in a flight with six others, and we were fortunate enough to get F4U's (Corsairs). We have done some checking in and will continue tomorrow. A number of my old friends are here; P—— has wrecked three planes and killed one man in the six weeks he's been here. Without sounding a blue note, there are a lot of accidents here every day, many of them fatal, but most are due to pilot error and not poor equipment. . . .

G.C.S. is a very small, quiet, pretty place in the midst of pine forests, and you will have a good rest if you can stay there in March. Rooms are not to be had. I found McG——'s wife at home. They have a "good" place—two rooms in an attic with bed, stove, and table. I am very happy you are not living in such quarters. . . .

March 2, 1945

Dear Phyl,

Just got your letters of 27–28 February and am quite happy and content again. Of course a swell steak dinner didn't hurt any. The food we get for 70¢ a day is amazing, and the service wonderful. For breakfast you can have eggs any way you want them or hot cakes, always a soup at night, and unlimited milk and butter.

More pre-flight training today. We are going over and over the engineering of the F4U and all cockpit operation, in order to eliminate troubles there after we start flying. As it will have been two months since we last flew Navy planes, we are due to get a few hours refresher course in SNJ's before we start in on the Corsairs. Our instructor, a Marine Corps captain, has had plenty of combat experience and seems willing and able to give us "the word." . . .

For once I am thoroughly blessed when it comes to roommates. P—— is ideal and has no idiosyncrasies to make him the least bit obnoxious. We have maid service to make our beds, sweep and dust, can shave in our rooms, and have a very pleasant existence.

One of the oddest things around here is to see the boys flying Corsairs walking around in flight gear carrying sheath knives. These are not to kill alligators in case of forced landings, as we at once assumed, but rather as a safety measure in case the collapsible rafts in our seat packs should inflate while in flight. The way the F4U is built, it would force the stick forward and the pilot upward against his shoulder and belt straps with exceedingly dangerous and uncomfortable results. Hence the knives. I've sent home for my trusty bear knife and will be carrying it from now on out, I guess.

March 4, 1945

Dear Phyl,

. . . 2045 on a hot, still March day. The heat of the day is still partially with me in the form of a very mild sunburn. This I picked up while fishing this afternoon.

We slept late this morning, missed breakfast, and so went into GCS early for dinner. We were invited to Ens. and Mrs. McG——'s for dinner and told to arrive at 1100, so were ready for dinner then. When we arrived, no sign of dinner, and so Mac, V——, P—— and I hopped in a little rowboat, stripped to the waist, and went fishing in the river. Our luck was fair to poor, and we soon gave it up for something far more entertaining.

About 1300 we noticed quite a crowd gathering on the pier, so we investigated and found it to be a Negro "baptizing." Having heard much about it and never seen one, we immediately cast anchor and settled back to watch the festivities.

The first half hour was taken up by a combined chanting and singing led by the preacher. All of the principals of the event were clad, head to feet, in white. They were all massed on the edge of the river, clapping, singing, literally dancing with "the spirit," and constantly urged to greater feeling by the exhortations of the preacher, a black of limitless vocal powers. The chanting and singing was most effective, if plainly barbaric.

After all had been worked up to a state of considerable frenzy, they were led down to the river, and the preacher and his assistant went out into the water to waist depth and were followed by the two men to be baptized. Prior to this time the chanting had been made up of phrases like "He's a great Lord," "How we gonna find de Lord?" and "Lord come to dis place." It now changed to "Lead me to de water," and the converts sang this constantly throughout the remainder of the service.

There was a great shout as the first man went under, but it took the second to really bring down "the Spirit." This one, a very tall and lean darky, came up praising the Lord, dancing and shaking like one in a fit. The preacher and his aide could not restrain him in the least. These two men then escorted the dozen or so women and babies down to the baptismal spot. I'm afraid the kids found it just plain cold, but not so the ladies.

Two of the women were especially active. One, rather slender, came out into the water making odd gyrations with her arms as she came, and upon being brought out of the water immediately went into a dance, hugging the preacher and attendants, bouncing up and down in the water, and crying out in "spirit language," completely unintelligible. The other was a fat old Mammy who went out quietly but when brought up, threw out her arms, fell back upon the preacher and uttered the most piercing screams I've ever heard. This she kept up for a good minute before she could be led back to dry land. There were many other variations, running the entire gamut of uninhibited emotional outlet. It was a most amazing thing to witness and, in contrast to the people calmly fishing nearby and planes flying overhead, seemed almost unreal.

We finally had dinner (incidentally, my only meal today) about 1500. In all fairness I must say it was a good meal. . . .

The base has a couple of sailboats for use by officers, and so P——and I tried to get one to take the kids riding, but there just wasn't enough wind to do the job, so we gave that up and just lay around the dock for a couple of hours resting our eyes and bodies by watching the wide river with its green, distant shore, the billowy white clouds, and the blue sky above. This place is perfectly described by the word "pastoral." . . .

March 5, 1945

Dear Phyl,

Another hot day of ground school, and I'm very thankful it is the last. Tomorrow we start flying again, after two months' lay-off. We are due to get a dual area check-out in the SNJ, and a couple of solos, just to get our hands back in, before tackling the Corsair. We've had exhaustive cockpit check-outs and lectures on every phase of operation of the big plane and are all eager to get started on the big job.

Just finished getting my flight gear lined up—coveralls, newly washed summer helmet, new earphones—what the well-dressed Corsair pilot is wearing in GCS this spring. And I mean that's all. Pants and shirt are strictly out. With the coming of warmer weather, a station order now allows us to wear khaki or greys without blouses while on the base and before 1800. My little-worn khaki blouse gives a sharp two-tone effect with my very faded old cadet pants, but it sure beats greens on hot afternoons. I'll try to look more presentable when you arrive and am contemplating a set of light worsted greys if possible. . . .

March 7, 1945

Dear Phyl,

. . . Today I got my area check-out, which consisted of an hour's ride in the back seat of an SNJ, with Capt. Leslie, our flight instructor. We will have one or two more hops and then hit the Corsair. When not flying, this is practically a rest camp. The past couple of days I've spent the afternoons mostly sleeping and reading. Not a bad life at all, if one has to be in the Navy, live on base, and have his wife a thousand miles away.

March 8, 1945

Dear Folks,

Today was the big day, and the old Navy poster saying, "You Too Can Win Your Wings of Gold and Fly a Sleek New Corsair," came to life. Of course old 185 is anything but new, but it still is a Corsair, and flying the "Iron Beast" is one whale of a big thrill. Yesterday and today I had refresher hops in the SNJ, shot a few landings at St. Augustine, and got my solo in the U today.

To put it mildly, I am greatly impressed. 2000 H.P. up front and six tons of metal underneath is not a combination to play with. You have to fly the U every second to even keep even with it. The plane was built for a big man, and believe me, I have my hands full herding it along. I thought the SBD was quite a plane, but it was a pleasure craft compared to the U. I am happy to say I had no trouble landing it today and didn't forget a thing, but I've never worked so hard in one hour of flying in my life. Flying this plane is done as much with the left hand as anything, because you'd really wear yourself out if you didn't use trim tabs constantly. As far as speed goes, I saw 240 K straight and level today with only about half power. That's moving, and it takes no time at all to cover our entire area. You can go halfway across Florida in 15 minutes. The only trouble with flying here is the visibility. There is a constant haze that cuts it down quite unpleasantly. I'm not saying yet how I like the Corsair, because as yet it is still the boss when we are flying, but I'm anticipating mastering the plane, and when I do, I'll take no guff off any pilot in any plane, because this plane is tops in any man's air force.

Other news there is none. We live a very quiet life here. Movies and letter-writing at night, the rest of the time we are really earning that 50% flight pay. Had our first PT today. We have inter-flight competition, and we did pretty well in basketball. Yours truly made the last and winning goal for Flt. 476—but man, oh, man, this day has sure shown me what 34 days of leave can do to one's physical condition. . . .

F4U Chance-Vought Corsair flown in Operational training at Green
Cove Springs, Florida (Courtesy U.S. Naval Institute,
Library and Photographic Services)

March 11, 1945

Dear Phyl,

. . . Flying goes along well enough. I'm having no trouble landing
the plane, but acrobatics are rough. An entire afternoon of them really
wears me out, from the bottoms of my feet as a result of kicking rudders
at 250 knots, to the top of my head from inverted flight. This week
should see us finish acrobatics and also formation. Our ground training
is really good, and exactly the stuff we've been needing and wanting,
but not getting until now. One thing I like about the set-up here is that
we do not fly the day after liberty. This is called a "sober-up" day, and
is really appreciated by us all, because no one wants to fly right after a
big liberty. P.T. here is quite easy and uneager; although our flight is
most uneager, we have a pretty fair record so far, winning in basketball
and volleyball.

March 14, 1945

Dear Folks,

Things are going pretty quietly here at Lee Field. The past several
days have been lousy with fog or low clouds, and I haven't flown since

this time last week, so there's nothing new on that angle. However, I'm in no hurry to finish here, and the longer it takes, the better. We continue in ground school with much valuable material available and well presented. We are at last getting a decent ground gunnery course using sighting range and deflection devices, and also the Gunairstructor. I probably wrote about it from Kingsville—a device wherein you sit in a cockpit and fly, trying to hit a plane projected on a screen, which is controlled by the operator. It can be excellent training, and the enlisted personnel here really know their stuff. We are especially fortunate here, as this base has done much experimental work on such devices and is way ahead of most. The engineering lectures are endless and really informative. Our ACI (Air Combat Intelligence) is right up to date and has the best collection of reports I've yet seen. . . .

March 15, 1945

Dear Phyl,
    . . . I had a little fun today on our first hop in a week. Went up at 1030, flew around with P——— in section formation a while, and then we split up to practice acrobatics. At 1145 I let down near the base and made radio contact requesting pancake—that's permission to land. Well, at 1300 they finally let me come in after watching about 100 planes land and refuel and take off again. I not only missed noon chow, used 40 gallons of gas, got *very* mad at everything and everyone, but also by my apparent absence, scared my instructor and roommate out of a year's growth. They were about ready to send out a searching party by 1230, and at that time I, circling above the field, was about ready to say, "To hell with it," and bail out. If I hadn't been so hungry I'd have really used my rank and reamed out the Waves in the control tower who kept me up there an hour and 15 minutes. I was seeing red and breathing threats of using my knife (we all carry knives, here) on someone, the instructor was worrying himself sick about what might have happened to me, so the combination provided a good laugh for us all. After this, I'm not calling the tower but just coming on in and landing.

March 16, 1945

Dear Phyl,
    . . . Had a very interesting time this morning. Very little fog, and we got off the ground at 0830. Capt. Leslie, our instructor, was checking

four of the flight on their acrobatics, the rest of us were practicing them. By 0900 there was a heavy layer of what looked like clouds below me, but I thought nothing of it and continued practice. Then at 0915 I caught the tower broadcasting a general recall to all aircraft. I could still see the river, so I had no trouble finding the field and requested landing clearance. However by the time I had made one orbit of the field, it was completely socked in and landing was impossible. The tower was assigning sectors for each flight to orbit in, and I found another Crimson plane (our flight name) and we tried to locate the rest of the flight. After a little radio contact with Leslie and considerable chasing after the wrong flight, we joined up with them and Leslie took charge. He is a swell guy, has had two tours of combat, and immediately put us at ease about the situation. We throttled back to maximum endurance settings and after a while were instructed to proceed to and pancake at Mainside—Jax. We had no radio contact with the tower there, and as they have heavy traffic, had a sweet time coming in between PBY's, F4U's, SNB's, SBD's, C-47's, and the rest of the Lee Field planes. I was flying wing on Leslie and made a swell carrier landing approach, but on landing, my tail wheel blew out, and so I pulled up in front of NATS hangar, folded my wings, and smelled like ten dump fires as a result of the hot rubber.

This was at 1030. Twenty-one planes landed at Mainside, more at other fields north of Green Cove Springs. I soon got a new wheel installed, and by 1140 we were cleared for return to Lee Field. Hit the deck at 1210, very happy to be back and quite happy about a new and different experience. The fog which had appeared to be going out at take-off time had really come back with a vengeance, and the mechs here said you could hardly see from the line to the hangars at 0930. No one ran low on gas, but when we left Mainside, Leslie, as acting C.O. of the Lee Field flights, signed for 2100 gallons of gas!

March 20, 1945

Dear Phyl,

. . . Three flights today, formation, and a real work-out. When leading a formation or flying wing on a good leader, the F4U is a swell plane, but if you get stuck on the wing of some throttle-jockey it can really be hell in heaven.

We have a flight-party pot in our flight; each man pulling a "Dilbert," that is something dumb, contributes $1. Today was our first day and we netted $2. M—— paid the first buck for an uncontrolled take-off that got away from him and he headed cross-country. Mine was the second for a scraped wingtip on landing. We had a bad cross-wind and I dropped a wing on landing, but was able to prevent a ground-loop. First damage credited to me since my last wingtip at Primary. However I was recompensed by leading the second division on our last hop— always a tough job—and hearing Capt. Leslie say it was "a damn good job." My three wingmen flew a beautiful position, and I hope we'll be able to work with that group all the way through, i.e. myself, P——, C——, and H——, all good quiet boys who don't claim to be hot pilots.

March 21, 1945

Dear Phyl,

. . . We had a big day today, although only two hops. The first was an oxygen hop, up to about 20,000', the second was a gunnery hop— and making runs in this plane really takes a lot of work. Not that it is hard to fly, but the forces on your body at high-speed pull-outs are pretty great, and this is one plane that will take a lot more than the pilot will.

This evening there was an all-hands barbecue—all the roast pork and beef sandwiches, potato salad, slaw, olives, pickles, ice cream, cookies, beer, you could hold. Only trouble was we had a terrific wind, and it was really cold out for the first time since I've been here.

Here's a bit of good news. Student officers on this base will have liberty, from secure activities until commencement of activities, every night beginning next week. . . . It's the first time this policy has been adopted in Operational Training, and we are all very happy about it, as you can well imagine. We are really being treated like officers—and we love it.

March 23, 1945

Dear Phyl,

. . . Flying goes on apace. We are now on camera gunnery, beginning overhead runs. It's a lot of fun and a lot of work. You can really black yourself out in these planes, and it sure isn't funny. Not dangerous, though, if you realize what is happening. I'd never blacked out before yesterday, but I guess from now on it will be a fairly common occurrence.

March 24, 1945

Dear Folks,

I'm enclosing the picture of our flight taken when we first came down. The old Corsair is just as big as it looks, and is no small plane, regardless of what the bomber pilots may say. Flying goes on at a fair rate. We have completed our familiarization and formation practice work and begun on the first of the two tough assignments here— gunnery. At present we are doing camera gunnery, and every error really shows up on film. I'm learning at last what blackout is. Never had any trouble with it before, but now our speeds are great enough to put on plenty of "G's." It's a funny feeling, but not too dangerous as long as one is aware of what is actually happening, and there are numerous ways to combat it successfully. . . .

About carrier duty, the F4U is flown off CV's [carriers] and we will qualify for carrier landings here, although no telling where we'll make our actual first CV landing. They have a small CVE [escort carrier] here part of the time, but many go to Glendale, Ill., or Norfolk for that. . . .

March 29, 1945

Dear Folks,

It is a beautiful, quiet afternoon, and I am enjoying my day off just sitting around reading, doing a few errands I have no time for on work days, and relaxing in general. Probably go sailing later or perhaps shoot a little skeet. Speaking of the latter, last time we shot, I topped the flight with a 13 out of 15, which isn't too bad.

The last couple of days we have been flying off the field at Palatka, 25 miles south of here. Much more convenient. We fly the planes down in the morning, have two or three hops, then come back on a bus after eating noon chow there. We have had two firing hops, and I am happy to report that so far, so well. I have gotten 12 hits out of 300 rounds fired for a score of 4%. An average of 2½% is required to qualify. The

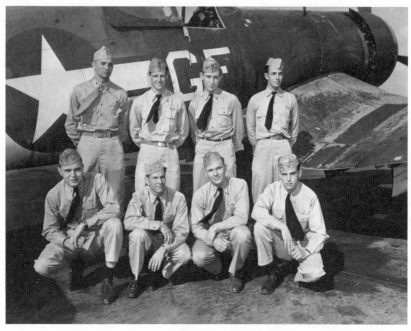

Corsair pilots, Green Cove Springs, March 1945—Captain Leslie, USMC, standing left, Ensign Rea kneeling second from left

average around here is probably 3%. When you are firing, a score of 4% looks pretty good, but when you figure that you made possibly 20 runs on the target, and considering the rate of fire, figure you were on target about one-half second, it turns out to be a pretty hit and miss proposition.

All our runs are the overhead type which diagrammatically look like this:

You start the run from an inverted position ahead of the target and get the lead. Then merely hold the lead and let the nose fall right on thru and pull out directly behind the target. It's a lot of fun, and a good run for hits, and is safe from possible opposing fire. . . .

April 10, 1945

Dear Folks,

I'm sorry to have delayed so long writing to you, but with Phyl here, my spare time usually allotted to letter-writing just disappears. It was wonderful to have her here, and we had quite a bit of time together, as last weekend was rainy and cloudy and we did not fly for three days. She stayed at the little hotel in town and had for company about a dozen Marine Corps wives, so it was pretty nice for her while I was busy on the base. We had dinner at the Student Officers' Mess every night save one, when the McG——s (roommate at C.C.) had us over for dinner. My day off, we went in to Jax, saw the town, and visited an old friend of Phyl's whose husband is at Camp Blanding, just west of Lee Field, a big basic training Army camp. They were very nice folks, both hailing from Richmond, Indiana. Phyl left Sunday evening, so I am now catching up with my letter-writing, and it's quite a job. . . .

Flying goes on much as usual. Light some days—as much as six hours others. My gunnery score is still safely above the line, and if we ever fly gunnery hops exclusively for a few days, it should be little trouble to get my required 75 hits out of 3,000 rounds. At present I have 18 hits out of 500 rounds fired. All of us are rather sad about losing our instructor, Capt. Leslie. He was transferred from USMCR to USMC, and as a promotion he was given the job of Engineering and Division officer for all F4U flights, which took him out of instructing. Our new instructor, a 1st Lt. Davis, is pretty poor in comparison, although he undoubtedly is competent. Leslie isn't too happy about the change either, and is working it so that he can fly with us part of the time, at least. Incidentally, Dad, the "army guy" in the picture is a Marine—and the Marines can fly—especially that one. I don't know if I told you, but Leslie was shot down off Guadalcanal on his first tour, flying SBD's, and spent 28 days on the island back of the Jap lines.

Last week a group of F7F's, the new twin-engine Navy fighter, came in here at Lee Field, and we got a good look at them. They are really swell—lots of fire power and speed, and the best arranged cockpit I've ever seen. One of the chief gripes about fighter planes is the complexity and poor arrangement of cockpit controls, and the F7F is really the first Navy plane to show any signs of the designers taking the pilot into consideration. Their visit inspired a lot of the fellows with a desire to fly them—naturally, and so they signed up for night fighters—about the only chance of getting F7's. There are a lot of good points to night fighters, but I'll take my chances in the daylight and take the F4U-4 against anything. . . .

April 10, 1945

Dear Phyl,

. . . Monday we flew two gunnery hops, one of which I did the towing for. I managed to keep my average up on the other one, so I feel that if we ever start flying gunnery hops with any regularity, I can get pretty good.

Today was our day off, so while the boys went in to Jax or St. Augustine, I stayed on the base, slept late, wrote six letters, read two books, and went to the show—"God Is My Co-Pilot." A pretty good show, in many ways far superior to most of the romantic aviation pictures. One thing really made it worthwhile. The hero and a priest are flying thru a storm and the going is rough. The priest recites a very good poem on the presence of God inside man, ending on His plea to men to "believe," and just as he speaks the word they break thru the clouds into the sunlight and virgin white clouds—"and they believed." It was very good from any viewpoint, and if you've flown in a cloud and known the clutching, smothering sense of powerlessness, almost of fear, and then the breathtaking thrill of breaking thru and seeing clear blue sky, blinding sunlight, and fairy castles of soft cloud—well, you know that there is more in life than meets man's eye on the ground.

April 12, 1945

Dear Phyl,

Another quiet day—no airplanes and no decent flying weather. To make it worse, Capt. Leslie could and would have flown with us today, had we been able. Yesterday we had one gunnery hop with Lt. Davis and it was strictly fouled up. With him along we just don't seem to be able to do anything right. I never saw a worse sleeve than the one we brought back—not 20 hits in it out of 1200 rounds fired at it. I'm still in the clear as far as average goes, but I'd sure like to get straightened out and get the qualifying hits out of the road in a hurry. Our plane shortage at present is acute, to put it mildly. We have in our group of five flights, enough planes to put two flights in the air at once, which means a full day's flying only once in a week at best. But enough griping—if we're not out until late in May, I'll not be sorry.

Tonight we got the news of the death of Pres. Roosevelt. The effect on the fellows was amazing. Many had cried out against him incessantly, but not a man but feels we have suffered a great loss. And as for the new president—well, most of us would just as soon, or rather, have the muchly condemned "Eleanor" than our new president. We are hoping it will make little difference in the progress of the war, but it looks like a terrific blow to the peace settlement. We can't quite see Truman dealing with such men as Churchill and Stalin and coming out with even his pants left. And the next three and a half years will see the big settlement made for the world in which we must live and work. It doesn't look good, and the joking prayers of last November—"God help the United States if Roosevelt dies"—are very real and serious tonight.

April 13, 1945

Dear Folks,

After a couple of days of foggy mornings we had one more gunnery hop today. Capt. Leslie flew with us, to our great satisfaction, and the hop went off like clockwork. Everyone shot well, including myself who, with only one gun firing, shot a 9%. Flying with Leslie we really click, and the mental and morale lift is really wonderful. Everyone gets along together and we get the job done in a hurry. We got our rounds fired today in just half the time usually required. The Captain will fly with us at least three more days, so we are hoping for good weather and a chance to finish gunnery before our new instructor returns from a checkout on Fuzzy. "Fuzzy" is the escort carrier "Guadalcanal" used to qualify pilots in this area for carrier landings. You'll probably hear more about her in a month or so. . . .

We were all deeply shocked at the passing of the late president. And the concern with which every man on the station views the prospect of the next three and a half years is a reflection on the political intelligence of the American people. We all hope that events will prove our worries needless, but the prayers of November, "God help the U.S. if Roosevelt dies," are no laughing matter now. As you know, colors will be flown at half-mast on all naval stations for a month, and tomorrow morning memorial services are to be held. At least the progress of the war in Europe is most encouraging, and we are hoping that the end of this year will see the army switching its power to the west. All new army groups are being sent to the Pacific now, but the end must come in Europe before any great drives are begun.

I get amusing letters from Jack W—— telling of his adventures as lord and master of a dozen or so Nazi POW's in France. T—— is working hard but liking it in K.C., and planning for an August wedding. . . .

April 14, 1945

Dear Phyl,

. . . We are now halfway through our gunnery course. My average is still aboveboard, but I'm hoping for a good day to really put it way up. Actually all I need to do is shoot a 1¾% on the remaining targets, but I'm hoping to do a lot better than that. S—— and M—— have really gotten hot the last two days and are way ahead of the rest of us, but the whole flight is well above the average.

April 15, 1945

Dear Phyl,

Boy, how I wish you were here tonight. . . . Oh yes, we're celebrating the fact that your husband got hot in gunnery today and fired 300 rounds for 34 hits—26 of them on the last 200, which means I have qualified in gunnery now, and all future hits go to raise my percentage above the required 2.5%.

April 17, 1945

Dear Phyl,

Just time for a line between chow and night-flying. Finished our slow-tow gunnery hops today. So far I have 81 hits for 1575 rounds, which is about 5 + %.

April 17, 1945

Dear Folks,

Very little news from Florida tonight. Flying goes on slowly but surely. We are two-thirds of the way thru gunnery and are starting to shoot at the high speed tows, which means our averages will suffer. I have plenty of hits, however, and a 5.24% average at present, so I'm sure to qualify, in fact already have. Got hot Sunday and hit 26 for 200, my best so far. Think I got quite a few hits today, but we only got back half a sleeve, so there was no way of telling.

We were supposed to start night flying tonight, but a little thunderstorm came up obligingly at 1930, and we didn't have to fly. One of the boys from our old Kingsville flight got lost last night and spent half an hour shooting touch and go's [landings] at Cecil Field, some 20 miles north of Lee. The poor guy didn't know he was on the wrong field until he saw some SB2C's in the traffic pattern along with his lone Corsair.
. . .

We will probably be here a month more at most, and it looks like we'll be heading west for San Diego—no East Coast orders available. If so, I'll have no leave, but enough time to spend a couple of days in Wichita, I think. It's all guess, so don't plan on it. Be sure to let me know when Bill Roy comes down here, even if you don't have his home address—the military address will be just as good. . . .

April 20, 1945

Dear Phyl,

This night flying is raising hell with my letter writing. . . . Flew last night two hops, touch and go landings, and just flying around. The old U is a lot better plane for night flying than I expected, so I'm quite pleased with the prospects of the remaining three or four nights of flying.

We've been having lots of clouds this week, and our flying has moved slowly. Yesterday afternoon I got in one combat hop. The day before, we had a GCI hop, that is "ground controlled interception." There were clouds of all kinds from 3,000 feet on up to 25,000 feet. Our division went 50 miles south, up to around 20,000 feet, and just circled, as the other group never was able to get thru the intervening clouds. These old planes aren't in condition for high altitude work, and at 19,500 feet my engine and C——'s cut out and we were barely able to maintain altitude at 100 knots. Visibility was pretty sad last night, too. You could barely see the river from 2500 feet.

Everyone got a big kick out of the cartoon in reference to my moustache. It is progressing fairly well, but has a distressing number of blond hairs in it! Guess I'll have to use a little shoe polish on it. H—— signed up for night fighters, leaving only 3 of us for day work, and the rest of the boys are putting the pressure on us to go with them to Kingsville—but unsuccessfully. . . .

This afternoon we are scheduled for a couple of gunnery hops on the fast sleeve, towed by a B-26. However, with the plane shortage, no telling what we'll do. We are flying indiscriminately with Leslie or Davis, and have gotten on to how to like Davis. The secret is just to ride him as hard as he rides you—take no guff from him and he's a pretty good egg.


April 22, 1945

Dear Folks,

I can remember a time, not too many months ago, when I used to hope I'd find some place in the Navy where they let a guy fly as much as he wanted to. Well, this has certainly been that place the past week. From the 19th thru the 21st, we flew 21 hours, and of this, 8 hours were night hops. We've had just about every kind of hop, regular gunnery, high-speed gunnery, tactics, ground-controlled interception, bombing, strafing, night bounce hops, and night section hops. Spare time just hasn't existed for the past three days. Consequently, all of us were dog-tired by the last hop yesterday and are spending our day-off catching up on sleep.

The GCI hop was very interesting. It was too cloudy to work out an interception, but we had a good altitude hop out of it. The division I was in went south 50 miles and got out from under the lower cloud layers at 3,000 feet and climbed up toward the tops, which we estimated at 30,000 feet. We had oxygen gear, of course, and got up to about 25,000 feet before our planes gave out on us.

The chief difficulty in high altitude flying is electrical, and when the engine cuts out you can't do anything about it but go back down. The plane I had cut out a little above 20,000, and so I had to circle at 18,000 at about 100 K until we headed back toward the base.

Night flying in the Corsair is much better than I thought it would be, and except for the late hours and the fact that we had already flown a full day prior to night flying, it was a lot of fun. I couldn't see just circling the base for two hours, so we flew a little DR navigation from GCS to St. Augustine, west to Palatka and Gainesville, north to Camp Blanding, and back east to G.C.S. This was a navigation problem in that the haze at night around here limits visibility to about 10 miles. . . .

April 22, 1945

Dear Phyl,

Sunday, evening, complete with a cooling rain shower, to top off a day of rest. And I really needed it. In the last three days we have flown 21 hours, considering that 8 hours of it was logged after 2000, that's quite a lot of time. It includes nearly everything—gunnery, ground controlled interception, tactics, high-altitude combat, instruments, dive bombing, strafing, night bounce hops, and night section. I used to wish for a time when we would fly long hours, and now that wish has certainly come true. Saturday none of us were in decent shape to fly for loss of sleep, and this morning I slept until 11 o'clock. We are rapidly completing the course here, and another week should see us into the last phase—field carrier landing practice, commonly known as "FCLP." We are scheduled to leave here about 12 May.

April 28, 1945

Dear Folks,

A quiet Saturday afternoon, the kind of a day you'd like to lie around the house and read and listen to the opera. Anyway, it is our day off and we are catching up on the relaxation.

Last night we had our flight party at St. Augustine. We had it in the beach cottage of one of the instructors and had a most enjoyable time. There was quite a crowd—two flights, three instructors, and assorted wives and dates. We got off early Friday afternoon and spent a few hours looking over St. Augustine. It's a quaint old town but as completely and thoroughly set for tourist trade as any place I've ever seen. Everything is "the oldest in the United States" and admission is "quite reasonable." The party itself was a gala affair of hair-letting-down, and when you get a bunch of old pilots like Leslie, Davis, and Davies together, you hear all the stories that have come out of the early South Pacific campaigns.

Flying is going slowly but steadily. We still have a few days' flying before we start field carrier landing practice, and one more night hop, but we're pretty apt to begin FCLP this coming week. . . .

May 2, 1945

Dear Phyl,

. . . Tomorrow will conclude our regular course here and we will begin carrier landing practice Saturday after liberty. Our last night hop was a very long one but went satisfactorily. Funny thing, you're tired the day after a night hop, but it isn't until the next day that it really hits you.

May 4, 1945

Dear Phyl,

. . . Yesterday we finished the regular flight course and are ready to start the low and slow flying tomorrow. Our last hop was dive angle estimation practice for rocket work. I still don't see how you're going to get a plane to do 400 miles an hour straight and level; in a seven thousand foot dive with partial power we were barely hitting 400 m.p.h. when we pulled out. That's a good speed, but pulling out of a long dive is sure a lot easier on a man's insides than a sharp pull-out from a gunnery run. . . .

The news from Europe could not be better, so I am hoping, as are all of us, that we can soon turn our entire effort to the west and bring war and uncertainty to an end. No, it will not end uncertainty, but it will allow us to lead the life of civilized men once more.

May 4, 1945

Dear Folks,

Another busy week over, and it is now 1130 and I'm just beginning the day, having spent the morning lying in bed with the *Reader's Digest*. In the May issue you'll find a couple of good articles on Naval Aviation. The one on the G Suit mentions Lt. Com. E. Scott McCuskey—he's our Asst. CFI [Chief Flight Instructor] and the best tactician and one of the best and most liked pilots in the Navy. The other, "Mission Beyond Darkness," is a pretty good description of CV [carrier] action. We have several fellows here who were in on that raid.

Yesterday we finished the regular course here and are now ready to begin carrier practice FCLP and CQ [Carrier Qualification]. This will take at least two weeks, and more, I hope. Had a hop yesterday on glide angle estimation. Dives of 40° starting at 8,000', pulling out around 1500'. Using a small amount of power, it is no trouble to hit 400 m.p.h. The next two weeks will be quite a contrast as carrier approaches are made at 80–85 knots. . . .

May 7, 1945

Dear Phyl,

. . . I realize I'm supposed to have left Lee Field, but I was held up from going out on the carrier for a couple of days and expect to go out any day, or something. . . . Today we watched FCLP again and had a dual in the SNJ. It's going to be a rough go and demand all possible hard work and concentration—and I'm not being pessimistic. V—— had his last FCLP hop today and failed to qualify, so is getting extra time along with three others in his flight. We have a bunch of hard-nosed landing signal officers (LSO) and before they give you a cut your approach has to really be perfect. The requirements are 8 out of 8, that is eight cuts out of eight consecutive passes. That means real flying. It's the old joke about "mother's advice to fly low and slow" come true. Maximum altitude is 150 feet and speed 85 to 73 knots. That isn't even fast in a Stearman.

The big noise today about V-E. Perhaps we all feel a bit cynical, but it seems to all of us that the public is making a lot out of the mere beginning. Look at a world map and compare the size of the late ETO and the Asiatic theater, consider that two-thirds of the Jap army has never seen a white soldier and that already the Japs are using drastic suicide measures, with gas a constant and probable threat, and there is much to take with great seriousness.

May 11, 1945

Dear Phyl,

   ... We've had 3 hops in FCLP and I guess I'm getting on O.K. It's pretty interesting, if hard work. The worst part is just sitting around waiting to fly. The old Corsair handles well enough, but every pass must be perfect after four more hops, and that's not the easiest thing in the world. The half-day is due to the completion of ground school. Just flying from now on.

May 11, 1945

Dear Folks,

   It's a rainy evening, and I am catching up with my correspondence after a swell liberty in St. Augustine. According to pre-arranged plans, I went in Wed. evening, and Bill arranged for me to stay at the same camp with them, so we had quite a time. Big dinner that night. Found a place that sells steaks as long as your forearm, so you can imagine how well and long we ate. Slept late the next morning and went out on the beach for several hours. In the afternoon we went thru old Ft. Marion and rode around town in a surrey, seeing all the oldest houses and stuff. . . . Big seafood dinner and show that night wound up festivities. Bill and I got the last several years well talked over, and I got acquainted with his wife, whom I like very much. The ocean was perfect for swimming Thursday morning, so the whole thing was a great success.

   Flying goes so slowly. We've had three FCLP hops so far. It's interesting work and very exacting. You have to be perfect from beginning to end, and once you get off, you might as well go around again. I'm getting along O.K. so far, just hope I continue to do so. . . .

May 16, 1945

Dear Folks,

A very quiet day off. Hot, cloudy, and sultry, the kind of a day you don't mind taking it easy. Flying goes as slowly as ever. We have had six FCLP hops and are all sweating it out. None of us has really gotten the word on it, but we're all hoping and trying everything possible to get it done.

Phyllis is due to arrive here tomorrow, having said farewell to Maine on the sixteenth. I have lined up a room in a private home which I think will be much nicer and a lot less expensive than living in a hotel, especially in view of the low quality of the Green Cove Springs hotels.

We will probably be here a couple of weeks more, at least, and then, pending satisfactory results in FCLP and CQ, be heading for San Diego. In the meantime, I wish you would find out for me the fastest train schedules and fares from Wichita to the coast, as I'll have no leave and every minute saved will be that much more time at home. I'll have eight days to get out there, and figure that six traveling should do it. Let me know what you find out. . . .

Been getting a lot of sun lately and have a pretty healthy tan, if I do say so myself. Out at Switzerland Field, where we shoot FCLP, there is no shade of any kind, and you just can't help getting a tan. When not flying, the standard uniform is a pair of gym shorts, so there is plenty of exposure. . . .

May 20, 1945

Dear Folks,

Everything lined up at the base, so late Sunday morning finds us enjoying a quiet, sunny period of letter-writing. Yesterday I finished field carrier with 7 O.K. passes, which is pretty good. Went out this morning and got my papers lined up transferring from the OTU (operational training) to CQTU (carrier qualification). No flying today, and tomorrow I'm slated to go over to Fleming's Island and catch a wire— that is, make four arrested landings as on a carrier. Won't go out to Fuzzy before Tuesday, and possibly not until later, I hope.

Phyllis got in Thursday morning, and we quite fortunately got a

USS *Guadalcanal,* CVE 60—carrier landing qualifications flown on "Fuzzy,"
May 22–26, 1945

Corsair about to "catch a wire" aboard *Guadalcanal*

nice little room to stay in. Very quiet and inexpensive. The weather here
has been ideal for the past week. Last night we were invited over to
McG——'s for dinner and had a great time eating corn on the cob and
steak. Very filling, very good.

[Continued by Phyl] Now it's my turn. This has all the earmarks of
being an extra special day. Bob went out to the base at 7 this morning
and at 9:30 he was back again. That is just the best way to start a day
that I can imagine!

This is a lovely, lazy little town. Nearly everyone is attached to the
Navy or the Marines in some way and so life is regulated by the base.
There are literally hundreds of Navy wives and babies and very few real
natives.

I still find palm trees and the ghostly Spanish moss thrilling, and
so when Bob isn't here I wander around looking and looking. Down on
the wharfs (or docks or whatever they are) you can always find little
darkies fishing. There are beautiful lavender water lillies in great patches
over the water. All in all there is plenty to see and enjoy. . . .

May 27, 1945

[From Phyl]
Dear Mother and Dad,

Bob returned at last after five . . . days and nights. . . . He tells
wonderful tales of life aboard a carrier. And really feels a great deal
better about being carrier based now. . . .

Right now he has gone to the base to find out when and how we
are to leave. I hope some leave will be given, but it's still just a hope.

We're pretty lucky, really, to be on this side of Kansas, because
going to California means we must go thru Kansas—and oh! we're both
so anxious to see you. . . .

The only drawback to this life lies in the constant packing and
unpacking. At present I am in a fine state of wrinkles. Men's clothes
certainly stand packing far better than mine. . . .

[Continued by Bob] Have arrived and will continue. Much to tell
about the past five days' sea duty, but will wait until we can talk it over
at home. Expect to get home latter part of this week. Arrival date
indefinite, but not too important as I have a 15 day leave! 'Til then, I
remain your qualified fleet pilot.

# 8

# War's End

**W**HILE BOB AND PHYL were en route from Florida to California, in Europe there was peace, the sad phenomenon of displaced persons, and the hunt for war criminals, who would be brought to trial beginning in the fall. Hitler, who had killed himself, and Mussolini, slain by vengeful partisans, had already been condemned at the bar of human history. American troops who were not assigned as occupation forces were being shipped back to the United States, where many boarded troop trains bound for West Coast ports of embarkation.

Out in the far Pacific, an invasion force of massive size was being assembled, even as B-29s swarmed over Japan and carrier planes swept in to hit airfields and other military targets. Ensign Rea had every reason to believe that he would be part of the invasion force which would face Japanese home defenses numbering three million men, hundreds of kamikazes, and a desperate, psyched-up population that was dedicated to victory or death. The weight of American preparations left no doubt as to the outcome, only as to the cost.

Three weeks after Bob and Phyl arrived in California, the predawn darkness of the New Mexico desert was lit by a flash brighter than the sun at noon; it disclosed an ominous mushroom-shaped cloud that surged defiantly toward the heavens. Few had any inkling of the event or its impact upon the future, certainly not a young naval aviator involved in the last phase of his stateside training. For him the summer of 1945 began with "the last great train ride" from Wichita to Los Angeles. In those days there were still Indians dancing on the railroad station platform "out West" and porters to prepare the berths in a Pullman compartment at night. In retrospect, it was a historic experience, never to be repeated in later years, as air transport replaced railroads in the United States. Ensign and Mrs. Rea were blessed with the hospitality of their relatives, Pearl and Stanley Van Winkle, who provided them headquarters in Hollywood for the base-hopping that ensued—a priceless advantage when Los Angeles motels were renting rooms by the hour.

Eventually the Reas and VBF-98 landed at Thermal, an appropriate name for a desert crossroads. The flying was potentially interesting but frequently frustrating. Pilots were introduced to the gadgetry of airborne radar but not to its use. A few rockets were fired—and all too many refused to leave the wing launchers. Flying the Navy's hottest plane, Corsair pilots were told to lower their dive flaps and brakes and slow down, for they would likely serve as glide bombers in the Pacific. Squadron exercises were generally pointless and just escaped being catastrophic on one occasion when the senior officer leading the fighters dove his Corsairs through a flight of unsuspecting torpedo bombers. These "group gropes" made men wonder just how we were fighting the war—and winning.

Los Alamitos was a proper base in every respect. It provided practical training—at least the squadron could actually drop bombs on a few rocks and strafe dye markers out in the Pacific. The practice might come in handy. Night carrier qualification was the grand finale of the training program. At night the deck of USS *Ranger* appeared very small and its exact location under the protruding nose of a Corsair very uncertain. Nor did the Landing Signals Officer help much. As he put it, it was up to him to maneuver a pilot to the approximately correct position behind and above the flight deck; from there on it was up to God. That technique did not work very well, and AG-98 effectively put the *Ranger* out of commission in a short time. If the skipper lost his temper, the pilots suffered considerable loss of faith in the system.

Flaws in procedure also appeared on the unhappy occasion of the flight's midair collision. It proved impossible to make radio contact with the base, thereby delaying rescue operations for nearly an hour. There was little satisfaction in being told, the next day, that "it should have worked." When a man's life may be on the line, one wants to know that it will work. Such episodes raised questions about combat conditions that were not pleasant to contemplate. But then the truth about war seldom is.

The strange, awful, welcome news of Hiroshima and the atomic bomb was heard over a car radio on a crowded California coastal road. No one doubted that the war was over. Demobilization was inevitably chaotic, but historically it was accomplished quickly and smoothly. Civilian attitudes emerged abruptly—for where, in civilian life, are gray gloves required for inspection? Armistice Day, as it used to be known in memory of another war and another generation of Americans, was an appropriate day to begin the trip home. Home was not exactly the same, nor were we, but we had a better idea of what it meant and what it was worth.

# Squadron Training, Air Group 98, Southern California, June 22–November 11, 1945

San Diego, Calif.
June 22, 1945

Dear Folks,

[From Phyl] . . . Uncle Stanley phoned the minister of University Christian Church, here in San Diego, and he, a Rev. Charles Severens, got a lovely room for us. It is a large house quite a ways from town and in a very nice residential section. A young Navy couple live here with the girl's father—and have 4 extra bedrooms—so we fitted in quite well. . . .

[From Bob] Well, first, the trip out was pretty good. We had fine accommodations in a most modern sleeper. The only drawback was food. Standing in line an hour and a half for breakfast and dinner was the usual custom. Lunch was not served, so we grabbed sandwiches and cookies at the station in Albuquerque and San Bernardino. It was a long trip, all right, and we saw much of New Mexico and Arizona that looked just like it did ten years ago. . . . We were only four hours late, which wasn't too bad, and made perfect connections with the Van Winkles up in Hollywood. The visit with them was all too short, and I'm looking forward to seeing more of them.

So we came down here today, and all misgivings worked out beautifully. For once the Navy has told me, and here's the dope. I'll be at Los Alamitos until 1 Sept., taking fighter-bomber training in F4U's, then report to ComFairPac and on to Saipan. Six weeks to two months there, and on out to the fleet. That's the story and it's about as good as I could hope for. I'll go out as a replacement, but that will give me the chance to fly with pilots who really know the tricks, and it looks advantageous from here. . . .

Oxnard, Cal.
June 24, 1945

Dear Folks,

Never a dull moment—that's our motto, so, true to form, I'm someplace else than I expected to be. Instead of going directly to Los Alamitos, we were sent up here to the Oxnard field, about sixty miles north of L.A. It's a small base and nothing to look at. No one knows why we're here, and we are supposed to leave about Saturday for Los Alamitos. We'll do no flying here as all they have is a few TBF's.

Leaving San Diego was a pleasure. The base there is big and spread out to about the size of central Wichita. Riding a bus from BOQ to the Ad[ministration] Building is comparable to going from home to the warehouse in Wichita, and I must have ridden buses 15 miles getting checked out there. That's not including the boat trip from the island to shore.

We had a good time in San D., however. Secured a room through the pastor of the Christian Church—good friend of the Van Winkles, so had a place to stay. Travel from San Diego to L.A. is merely a matter of getting to the station an hour or so before train time and getting in line early, so not too difficult. We spent Saturday night with the Van W's and I came on up on the noon train. Phyl is staying there at present while we hunt a permanent place to stay near Los Alamitos. Pearl and Stanley have been wonderful to me, and it really means a lot out here to have some sort of headquarters. . . .

Long Beach, Calif.
July 2, 1945

Dear Folks,

[From Phyl] My, but life in a suitcase can get complicated! But it's fun, too. And I am getting quite adept at packing in ten minutes flat—or not unpacking, however you want to look at it.

Bob wrote when we first got to Oxnard, I believe. Well—that was quite an interesting little place. We were fortunate enough to find a place at a little motor hotel just two doors from a couple we had known in Florida. Joe—the Ensign—had a car, which has proved very fortunate for us—so he and Bob rode to and from the base together. Joe's wife Pearl is an awfully sweet little thing, and we had a fine time together. The days weren't half as long, when Pearl and I could talk together, as they would have been alone. . . .

Now—here in another motel in Long Beach we have our alarm set for 6—and then off we go again—this time to the desert. Luckily, Joe and Pearl have asked us to drive up with them—we also rode down with them from Oxnard. . . .

[From Bob] It's all very well to see Southern California, but from a Naval standpoint it's not so good. This thing of hitting four bases in ten days is not my idea of how to win the war.

Today, and today only, we were stationed at Los Alamitos, just a few miles from Long Beach. Now we're headed for a crossroads known as Thermal, out in the desert. Length of stay unknown, but this can be said for it, we are sure to see the sun for a change and temperatures of 135° are promised. I'm sending my blues home!

I haven't done anything of note in the past week. Just piddled around with some ground school and P.T. . . .

. . . Dad, if you can work it and not miss it too much, could you send me a couple of gas [rationing] stamps? Joe's car has meant everything to us the past and coming days, and while I can take care of the financial end of things, I can't help there. Really appreciate it if possible; if not, thanks anyway. . . .

Indio, Calif.
July 4, 1945

[Postcard]
Dear Folks,
    We are pretty well located again—living at Indio. The base isn't too bad, but the temperature is great—even hotter than Texas ever thought of being. We are due for a month here, by which time we should have regained all of our Florida tan and a lot more. Typical scenery—

Thermal, Calif.
July 8, 1945

Dear Folks,
    At last a day off and a chance to write. So here's the word on this place. The base at Thermal is an auxiliary to an auxiliary air station and is pretty sad as far as looks and organization goes. It is right in the desert and just a few miles north of the Salton Sea. Our elevation is −150′, with surrounding mountains at 5000′–8000′. The valley—Coachella Valley—is quite a date-growing center, but that is all you can say for it. The average temperature in the daytime runs about 120° in the shade, and by mid-afternoon up to 135°.
    As a result of this temperature, operations on the field start at 0300 and are secured about 1300, as the mechs just can't service planes when it's that hot. This means that the first hop is practically a night hop, and so gets rather well fouled up. The field has only two runways, but as the wind is almost always steady North or Northwest, it isn't too bad.
    We have flights of six men, and I am in with 5 from the same flight at Green Cove Springs. We've only had one formation hop, but it was very sad. Charley K——, who was my roommate at Kingsville, and Joe M——, whom we are traveling with, are the only other Corpus boys in the flight, the other three having graduated at Pensacola.
    Getting up at 0245 means hitting the sack early at night, and as it's too hot to go out in the afternoon, and nothing to do anyway, this is a quiet life. We are sharing a sort of duplex cabin at a very nice motor court in Indio, the nearest town of any size, with Joe and Pearl M——. It's somewhat inconveniently located, but so is everything out here.
    Yesterday we got ahold of a .50 calibre cartridge case with a metal

liner, so now we are having cold Cokes and ice water whenever we feel like it.

The planes at Thermal are quite a disappointment. Instead of new F4U-4's, we have F4U-1A's, same as at Green Cove, and they are in very poor shape. As a result we have to take a lot of ribbing from the F6F pilots whose new planes really can run circles around these old crates that aren't much better than good SNJ's. . . .

Incidentally, I haven't seen a letter since we left home, so how about a special edition right away. . . . Also, please send with first letter my CAA license, if you haven't already. There is a little commercial field here, and outside of one picture show, it's the only other form of entertainment.

July 12, 1945

Dear Folks,

[From Phyl] This is certainly a topsy-turvy life! Bob flies in the very early morning—3 a.m. Monday, 4:30 Tues., and so on! And so you can imagine how happy Bob is about getting up! He never has liked getting up in the morning. Of course this all means that bedtime comes at 5 p.m. or at least by 7. And that is far too early for me to be sleepy. Generally I tuck Bob in and then slip outside and chat with Pearl M——.

This is interesting climate—Bob says "interesting" is a gross exaggeration. 131° in the shade! Luckily we have a nice air-conditioned cabin. In fact, without air conditioning here I don't know how the people could stand it. Even cars are air-conditioned!

There is a great date ranch about 2 miles from here that is a regular resort. A beautiful home has been built there and presented to the Navy. The "Skipper" of the base lives there, and the officers and wives are allowed to go there to swim in the beautiful swimming pool.

Pearl and Joe and Bob and I have had lots of fun there. The boys have to put in an hour a day at some kind of exercise, of course. And that is the coolest way to do it. . . .

[From Bob] Between this schedule and the heat, my letter writing has definitely gone to seed but please excuse. The worst thing about these early morning hops is that I have to get up an hour and a half before the hop, which, in view of 0300 flights, makes the night pretty short. . . .

Time for my lunch and Phyl's breakfast—

July 13, 1945

Dear Folks,

Lots of mail the past couple of days, so I guess I'd best reciprocate. This is a good time, as I have a 24 hour duty watch to stand, have to stay on the base, and have only about ten hours of work. After six hours here, I am most happy that I'm living in town. There's no place here to cool off, and recreational facilities are about as limited as they can be.

I believe we've caught up with our mail now. . . . Got both the letters with gas stamps, and believe me, they are really appreciated. Joe's monthly ration is just enough to get us to and from the base, and now we may be able to get up to the cool air of sea level sometime we have a day off. You were most generous, and I hope you didn't cut yourselves off short.

The weather continues hot, and the only relief outside of air-conditioning is the Skipper's swimming pool. I'm having quite a time trying to teach Phyl the various strokes, Navy style. And she is also having a time trying to get a tan. . . . Myself, I am rapidly approaching the shade of our new bedroom suite without any effort, and the swimming is certainly good for me, as I've done little in the past year. . . .

Flying goes on slowly and dully. I am constantly amazed that these crates we have fly. No one on the base except the pilots would be caught in one, but we're always game for a try. Today was typical: a 0415 predawn hop in which our division failed to get joined up, followed by a rendezvous of all F4U flights for break-up. We had 4 divisions, and the Dilberts leading the hop stepped them down and then circled the field with the top division at 500 feet. With an ordinary interval of 100 feet between divisions, you can guess that there was some fancy flat-hatting. Personally, I don't go for division cross-unders at 200 feet, nor do most of us, and there was some real word-giving when we got down. Then on a gunnery hop, the tow-plane flew 15 knots slower than he was supposed to, and without tracer ammo you can't get hits at 90° angles with that set up. I was especially disgusted as I needed hits and my runs were about the best I've ever made.

All of which doesn't sound very enthusiastic, but I have my wife to keep my disposition sweet, so all works out pretty well. . . .

July 15, 1945

Dear Folks,
. . . Flying goes on apace. Today we flew from 0500–1500 with time out for lunch and ground school, each 30 minutes long. Not too bad except for the heat which makes a closed cockpit unbearable.

This morning I had the lead of VBF [the fighter-bomber squadron] on our pre-dawn strike, or "group grope" as we call it. Went off better than usual if I say so myself.

Also today we started the course in rockets. It's a lot of fun, diving in at 15°–30°–50°, but plenty difficult, as the slightest errors in speed, angle of dive, and deflection throw your hits way off target. Much more difficult than gunnery or bombing, but something new for a change and quite interesting. . . .

July 25, 1945

Dear Folks,
After a week and a half of steady flying we've finally gotten a day off, and it's pretty good to lie around a while and catch up with things. This place looks much more like home now with a few pictures around it. . . . The C.A.A. license came through a couple of days ago and is being put to use already. According to the new civilian pilot regulations, with my license I can fly any single engine land plane and may take along any passengers, free or for hire, and may give instruction. So, I got checked out Monday in a nice new Taylorcraft, a 65 H.P. wheel-control job, and after shooting a few landings took Phyl up for a short spin. Needless to say, she was quite thrilled and more anxious than ever that I teach her to fly. . . .

Our stay here nears its end. We have three more rocket hops and will do a little field carrier work here, leaving for Los Alamitos the 1st or 2nd. According to schedule, we should get our day qualifications in on the carrier about the 6th, so we'll be moving around again in the near future. . . .

P.S. If you need the Ration Book number to get Phyl's sugar, the book may be in right hand desk drawer. . . . Coupon enclosed.

Long Beach, Calif.
August 6, 1945

Dear Folks,

[From Phyl] There has been a long silence but we have been thinking about you a lot even if the mail doesn't show it. At first we thought we would wait until we had an address here in Long Beach, but there is no telling when that will be, so here is the news to date.

We are staying in a small hotel waiting for an opening in one of the motor courts, and that is just about all we know.

Bob and Joe leave at 7 in the morning and get in at 5–6 in the evening. Everything is a big improvement on Indio but still leaves a great deal to be desired. . . .

[From Bob] . . . It's really great to be back in civilization again, even though living conditions are not the most desirable. . . . I especially appreciate a good night's sleep and the good breakfasts and lunches available on the base. Regular hours really make a difference in your morale.

Our flying here has consisted so far of field carrier hops and formation hops. We all need as much work as we can get on FCLP, and it will be our main activity here. Next week we go out for day carrier qualifications again, and when we return will start immediately on night field carrier. No one is looking forward to that, and I'm certainly no exception.

We've had several hours of just flying around over this area. It's most interesting, too. Every kind of terrain—hills, valleys, plains, sea, islands, farm land, and the metropolitan area of Los Angeles. The weather is perfect, but visibility is usually poor—a great deal of fog and haze all day long. Most confusing factor is that here at Long Beach the coast runs in a general east-west direction instead of north-south as one would expect. The area is dotted with air fields in greater number than I ever imagined. From 10,000 feet on a clear day above Los Alamitos, one can see about ten different fields—army, navy, marine, and civilian, and the air traffic is quite heavy. Only thing we haven't run across are Army fighters, with whom we are all itching to match our planes and skill.

Wednesday the AAF had a big day and an exhibit at the Douglas Long Beach field, which we took in. Almost everything there—P-40, P-51, P-80 jet job, P-38, P-61 night fighter, A-26, B-24, B-25, B-17, B-29, C-47, C-54, C-46, Constellation, and a couple of old pre-1918 models. Today we got to look at the new FR-1, Navy semi-jet fighter, that is just coming out, and it looks like a beauty—small and clean, and with the jet power can really climb. Also gave myself a cockpit check-out in the F4U-4 today. It's essentially the same as our old F4U-1D's but has a bigger engine and the cockpit has been cleaned up and redesigned, making it a much nicer plane. We'll get to fly them before we leave, I hope. The planes here are in pretty fair shape. At least they have good paint jobs and are kept relatively free of cockpit dirt and grease, which makes it much nicer from the pilot's viewpoint.

Friday afternoon we were given a couple of field catapult shots. It was quite a lot of fun—just the opposite from an arrested landing. You get flying speed before you know it. Certainly better than trying to take off unassisted when there's no wind on the flight deck.

August 9, 1945

Dear Folks,

[From Phyl] Pearl and I just returned from the base. We had wonderful luck today—At 7 we went to the base with Bob and Joe and all ate breakfast. Then Pearl and I wandered around getting fingerprinted for our passes and making vital purchases (toothpaste and ink). Then after luncheon we watched the boys fly off for the carrier. They are to fly aboard in San Diego. It's to be a four-day cruise and Bob thinks it should be quite interesting. . . . They should be back by Tuesday and so we hope to have our celebration right on time. . . .

August 17, 1945

Dear Folks,

Well, vacation is over and it's back to work for the USN. And I do mean work. Here it is two days since I wrote the first line of this letter, and this is the first chance I've had to continue it.

To go back about a week. We had a very good four-day cruise on the USS *Ranger*. Being the first CV built as such by the Navy, it is small by present measures, but was a lot bigger than the *Guadalcanal*. All we had to do was shoot four landings, so our part of the cruise went smoothly. We also had SB2C, FM's, TBF's, and F6F's out at the same time, so there was plenty to watch when we weren't flying. It was really cold at sea, much to our surprise, and we wore clothing and flight gear all the time. I got to fly my plane down to San Diego, where it was transferred aboard ship. Ten of us flew down, and we had a great time landing at North Island. It is the busiest Navy field on the West Coast, and we had practically an instrument approach to make, due to low clouds and fog. On top of that, we had to land on the mat instead of on runways—all of us made very lousy landings. We flew back to Los Alamitos in a C-47 transport.

As soon as we got back we started in on night field carrier practice and have been at it ever since. It's more instrument flying, and none of us like it very much.

We had the 14th off, and so we went down to Balboa and spent several hours there looking at the sights and riding around the bay and inlets in a motor boat. Then on the way home we got the official peace news and so went downtown to watch the celebrations. Everyone was downtown and Long Beach really went wild. However it was a good-natured mob and quite different from that in Frisco. About 1800 we got the word that AG98 had the following two days off, so Joe and Pearl left for their home near Frisco, and Phyl and I had the next two days to ourselves. . . .

So, for the past two days we've been catching up with our flying, night and day. The cessation of hostilities will have little or no effect on our activities, and we all expect to ship out for Pearl Harbor the first of next month and probably be in the Navy at least six more months. . . .

We've been having pleasant temperatures, but lots of clouds, fog, haze, and some rain. Very lousy flying weather, especially for night flying, although we got over L.A. one clear night and the lights were really beautiful.

Holtville, Calif.
August 21, 1945

Dear Folks,

Nothing like seeing the state while I'm here. Yesterday, because of poor weather around Los Alamitos, the squadron was shipped down here to Holtville, which on your map of California is right near the Mexican border, about 20 miles from El Centro. The country around here is even more desert-ish than it was at Thermal, but otherwise it is much better. The base is well layed out, kept as clean as possible, quarters are good and food excellent. And as we are here for night bounce [landings], the fact that there is no haze, clouds, fog, a full moon, and no obstructions, makes it very nice as far as flying goes. Phyl is up at Long Beach, as we were flown down here and no cars were allowed. We'll be here about three days, then back to Los Alamitos and out for night qualifications on the *Ranger*.

Had one of those things happen Sunday that make you stop and think. We were out on a hop chasing tails, and the second and third men had a nasty mid-air collision. They were Bob F—— and my buddy Joe M——. Knocked a wing off Joe's plane, the engine off Bob's, and it subsequently started to burn. Both of the boys bailed out and landed on Catalina Island. The two of us remaining in the air spent the next 2½ hours directing their rescue. It was pretty rugged where they landed and it took quite a while to get Joe out. He apparently made a bad landing and broke both legs. They were just bringing him back to the Long Beach Hospital when we left for Holtville.

The end of hostilities will have no effect on our squadron. At present the training program continues as before, and no one expects any sizeable release of men in the near future. The Navy "point system" is quite a joke. According to it, fellows like myself would have 2⅓ years more before becoming eligible for discharge. I rather expect we'll be in until our enlistment is up in November of next year, and we expect to leave for Pearl Harbor next month. Not the way we might like it, perhaps, but with the Navy asking for 30,000 more USN officers for peace time, none of us are apt to be home soon.

Holtville, Calif.
August 21, 1945

Dear Phyl,

As the old saying goes, it never rains, it always pours, or something. Yesterday morning [at Los Alamitos] we had ground school as scheduled. This included shooting skeet, at which your husband excelled above eight others—our flight and a bunch of FM [Wildcat] pilots. I had just sat down to lunch when your call came.

After lunch I picked up a carton of Luckies—cigarette rationing is off now—and investigated the State Dept. deal unsuccessfully. Everyone was out for lunch. Picked up the mail and went up to the ready room about 1245 where I first learned about our jaunt over here. We had to wait until 1330 for a lecture from *your* friend, Com. Harris, on our attitude now that the war is over, etc. He ran down at 1350 and I jumped into F——'s V-8 and tore into town to get my stuff. Had to be back in time to take off on a transport at 1500. I made the round trip in 35 minutes, which I wouldn't recommend for you.

The trip out here was rather routine. Flew over Indio, Thermal, down over the Salton Sea, and so to the base here at Holtville. Got in about 1630 and promptly got our quarters and a line-up on the base, chowed, and got ready to fly.

Although we're almost on the Mexican border, well south of Thermal, and even more out in the desert than was Thermal, this isn't a bad place. It is well arranged, fairly clean, quarters are quite satisfactory. Q—— and I are rooming together. The field is very nice—even better than Los Alamitos.

The reason we came down here, of course, was to get in our night field carrier, which was impossible at Los Alamitos due to fog and haze. It is really different out here. Almost a full moon, none of the haze, good visibility, make night bounce much more enjoyable than it was at Alamitos. We are scheduled to stay about three days and go out to the carrier on the 25th. It's not the most ideal settlement for us, but can't be helped.


August 28, 1945

Dear Folks,

Well, back to more or less normal again, and time to write a letter. We finished night bounce out at Holtville very satisfactorily. The FM pilots had some trouble, but our boys were all hot. Got back to Alamitos in time to spend one evening with Phyl, see Joe, pack, and leave the next day for San Diego.

Back aboard the *Ranger* it was much like our first cruise. Had fog until 1600 the first day out, however, and had to rush like mad to get in our two day refresher landings. The night landings weren't at all bad—only had to make two. And it only took 3 passes for me to get two cuts. However I hit a little hard on the second one and broke my tail wheel—

a very common accident in carrier landings. Well, despite the fact that all of the Corsair pilots got qualified, all ran far from smoothly. In little over two hours we had the following difficulties—3 broken tail wheels, one no-flap landing, one plane sent back to the beach unable to lower his hook, one barrier crash (my boy Johnny Q—— came in hot and high, took out 3 barriers, one AA gun, scraped all the paint off the island, and didn't get a scratch!), and an FM pilot landed beyond the barriers and on top of another plane without either pilot getting touched. By this time the Skipper was boiling mad, secured all operations and returned to San Diego, refusing to fly any more pilots until they had had more bounce drill. To top it off, he kept us all aboard ship at San Diego for an extra 24 hours.

So that's all over and we are back to the old routine of strafing a dye slick out at sea for three hours a day. We're supposed to have a navigation problem tomorrow morning, but will undoubtedly be fogged in and won't fly at all. As for future plans, nothing but scuttlebutt that isn't worth repeating. . . .

Joe and Bob, my boys of the mid-air collision, are progressing nicely. Bob is about ready for release from the hospital, and Joe is getting down to the long grind of his 2½ months flat on his back. Not having his car for transportation has not proved much of a problem, as bus service is quite good here.

Really got a shock tonight. We had just finished dinner and were on our way to a show when we ran into Marvin S—— on the street. He had been looking for us all day and just missed us, and had only an hour and a half before catching a bus for San Diego. Believe me, we really crowded a lot of visiting into that 90 minutes. Having just missed him at Corpus and coming so close here, fate finally broke right and there wasn't a silent second. He is being attached to a CASU-aviation maintenance outfit at Diego and isn't sure where he'll be, but I sure hope we can see more of him. Did I tell you that Lt. (JG) E—— G—— is in the naval hospital here, too? He is an old FU man and former Alpha Kap. Convalescing from an automobile accident that really shook him up. . . .

Lots of pressure on us now to sign up for the regular Navy—especially men with good education backgrounds. Only 18 more years and I could retire with a $100 a month pension. The thing that gets me in all this is that after two years of fairly decent work and concentration, there is no demand for that which I really know and can do best. The idea of getting up to fully qualified and unlimited Naval Aviator and then just tossing it in the drink kind of goes against the grain. Then also the idea of tossing off a $300 a month job for the very questionable future outlook seems to me to warrant a second thought.

September 3, 1945

Dear Folks,

Had intended writing two nights ago, but so much has been happening I haven't had a chance. We'll start out with the flying. The usual routine has been touched up a bit. We got in one navigation hop out beyond Catalina for about 60 miles, and now are having instrument hops in the Corsair. One plane on the line has an orange-colored hatch, and with blue goggles you can't see a thing beyond your instruments. Of course we always have another pilot chasing to watch out for other planes, vertigo, etc., and it's quite valuable experience. Then I've also had two hops in the new F4U-4. While the outside looks like the old U-1D, the cockpit is entirely different, and it handles a lot different. It has a lot of power and speed. I clocked it at 370 m.p.h. straight and level without nosing over to gain speed, and at a low altitude where it is decidedly not at its best. Just for fun I got into a dog-fight with an FM and sure saw why those boys like their plane. He could out-turn me in any position or speed, the only defence I had was to dive and pull away or climb and pull away, turning to get at best a single head-on shot. But when I poured the coal on, I really left him in a cloud of dust.

Joe M—— is coming along fine, and all that remains now is for his bones to knit and he'll be O.K. Bob F—— is home on a three week leave at present, looking fine, but pretty nervous. E—— G——, former FU man who is out at the hospital is also coming along and should be up in a few weeks. . . .

Last week—out of a clear sky—we ran into Marvin S—— on the street. Had a short visit before he went on down to Brown Field. He had two days off, as did I over Labor Day, and so came up and we had a swell time showing him Long Beach. We both got a big kick out of seeing each other, and I know how much he appreciates having someone from home to go visit—viz: myself and Bill Roy down at Green Cove. He really looks swell and apparently will be around for some time to come. No word yet on what they'll do with us, of course. Took S—— out to see G—— at the hospital, which really pleased G——, he and Marvin played basketball together at Friends.

About whether a water landing would have been better for the boys when they jumped—for F——, who was relatively unhurt, most definitely—he would have been given medical attention two hours sooner had he let himself land at sea. For Joe, it's hard to say. He was pretty much out of his head and might have run into trouble in the water. . . .

September 10, 1945

Dear Folks,

. . . Not much Navy news of late, and most of it is not factual. We are now on a peace-time schedule, the working day running from 0800 until 1630, with Saturday afternoon and Sunday off. Regular weekly inspections in blues, and shortly only blues to be worn off the base. Latest word is that we may be here for a couple more months. Air Group 98 is eventually to be a pool for men awaiting discharge. On the latter subject, little new. My best bet is still Nov. 1946, as a recent AlNav states that all officers may request and receive discharge upon the expiration of their term of enlistment. The point system will be lowered, but only when and as the Navy sees fit—a matter of weeks or years.

Recent lectures on the operation of the peace-time Navy have caused a complete reversal of feeling in the squadron, and from an original 9:1 feeling of interest in remaining, it is now about 2:8. In line with my own civilian interests, I have written Pres. Young [of Friends University] concerning the attitude of graduate schools toward people like myself. . . .

You asked for Joe's address. . . . He will really appreciate a letter from you and the more fishing in it the better. The boys in our flight bought Joe a fly-tying kit, and he's beginning to produce some pretty good specimens.

September 14, 1945

Dear Folks,

Not much news—this is mainly an S.O.S. Beginning tomorrow we are having dress inspections every other Saturday morning. This week it's in working blues, but next time it will be in dress blues. That means blues with *grey gloves.* Therefore I must needs request that you institute a search among my belongings, and when you find them, send them along to me. . . . I suggest looking in the bottom of my big green seapack, as I'm reasonably sure that the gloves are there.

Pretty easy life these days. Thursday and today I played tennis and shot skeet or pistol, and that filled my "working day." We haven't been flying because of fog and low clouds in the mornings. . . .

September 23, 1945

Dear Folks,

. . . As far as news goes, there is none. Everything is just as it was last Sunday when we talked with you. We did very little flying this week, due to fog and low clouds lasting late in the day. This week we're flying mornings, so will undoubtedly fly even less. Bob F——, one of the boys in the accident, got back off leave and is now flying with us again.

Joe M—— is coming along fine, and his morale has improved immensely. Dad's letter really went over big with him, especially as mail is his big event each day.

The grey gloves came thru in fine time and will be worn for Saturday's inspection. This past week, inspection was in work uniform, followed by a line check on the planes by the pilots. Lots of dull routine stuff, but the shorter hours are most welcome. Have been working on my sadly neglected tennis game recently. . . . Big trouble is to find someone to play whose game is good enough to improve mine. Most of the pilots just can't play, and the boys around the station are too good for me. . . .

While no further news about our being discharged, I am looking forward to next year and have written Pres. Young and Dr. Mekeel concerning graduate schools. No answer as yet, but should be soon. In the meantime we are seeing how tight the budget can be comfortably drawn (note enclosed check). I can see that the next few years will be quite full, but the more work I've had on my hands, in the past, the better I've done it and the happier I've been, so it doesn't look bad from this distance.

September 26, 1945

Dear Folks,

. . . Things at the base go on as usual. Lots of fog these days, and it is getting quite cool in the early mornings. Things will be even slower out at the base after this week as we are going off the ground school schedule this Friday. However I can still get in periods in Link trainer and pistol range in my spare time. There is some talk of cross-country flights as soon as everyone has finished their carrier qualification, which may mean next month.

October 4, 1945

Dear Folks,

. . . Now here's the more or less ungarbled word on my future activities—subject to immediate change. Tomorrow I fly my last hop for the Navy. Next week I'll be transferred to Brown Field, about four miles north of the Mexican border to await the coming of 10 November. On that date I'll put in my application for release from active duty and it should come through in a couple more weeks. Nothing definite or certain, but that's the way it looks today. Yesterday it seemed quite different and may well change tomorrow. Anyway, I'm on my way toward getting out, and so I'll have to turn in all my flight gear. This includes the *box of gear* I packed and wrapped . . . and which is (or was left) atop the big trunk in our lot upstairs. Please send it along as soon as possible. . . .

October 8, 1945

Dear Folks,

. . . There has been no flying for us reserve officers since last Wednesday. Today, however, we were given just enough time to fill our necessary quota for October flight pay, so I had one short hop which was probably my last flight in an F4U and for the Navy.

Most of the time we just sit around wishing we could fly, talking about the days when we used to fly three or four hours, and discussing what it will be like on the "outside" and what our plans are. No further word has been passed concerning our leaving for Brown Field. When it does come, it will probably be one of these "report at once" deals, which will make everyone very unhappy.

October 14, 1945

Dear Folks,

This week's silence has been due to conditions at the base and not lack of good intentions on my part. Monday morning all of us Reservists were ordered to be packed and ready for Brown Field at 0800 Tuesday. So I rushed home, made the necessary arrangements, packed, and at 0800 Tuesday, was ready to go, with my gear all out at the base.

Well, Saturday morning the conditions were unchanged. Every morning this week I go out to Los Alamitos expecting to leave, and every afternoon I come back, fully expecting to leave the next day. All my clothes are at the base, and I carry razor and tooth-brush in my pockets. We've been doing nothing out there except mustering, shooting pistol, playing tennis, and waiting to secure. The way it looks now, we may not leave Alamitos at all, and then we may still leave tomorrow. . . .

October 21, 1945

Dear Folks,

. . . Little news from my end of the line. The past week has been the same old story. However next week the Navy is inaugurating a new program for us Reservists. First big change will be muster at 0830 instead of 0800. Then we'll probably have a couple of classes, P.T., and secure early. No details yet, but it will give us something definite to do.

The range has run out of targets so no pistol shooting last week. However the tennis games continue to fill nearly every morning. Joe M—— is coming along fine and may get out of his traction set-up and into a cast soon. Friday, George W—— and I spent a couple of hours freshening up our Spanish grammar. Most amusing and confusing. Hope I haven't forgotten as much history as I have Spanish in the past 3 years.

October 30, 1945

Dear Folks,

. . . Well, we were all set to have our papers go in to Diego day after tomorrow, when this morning a directive came out from ComFairWest effecting the freezing of all pilots, regardless of points, in Air Group 98. There are better than a thousand planes down at North Island that the Navy wants distributed all over the United States, and it looks like we're the boys who are going to do it. How long we'll have ferry duty is problematical, anywhere from two to six weeks, so there's no telling when I'll get out. Before Christmas, I hope.

November 4, 1945

Dear Folks,

. . . My naval status is still in doubt. Saturday we were "unfrozen," so there's a possibility of getting orders for release soon. At least I hope it won't be much longer before something develops of a definite nature.

*Ensign Rea was released from active duty November 11, 1945, and returned home to Wichita with Phyl on November 16. He remained in the Inactive Reserve and was promoted to Lieutenant (Junior Grade) as of February 1, 1947; his resignation was accepted and he was discharged on May 10, 1956.*

# A PERSONAL POSTSCRIPT

Phyl and Bob Rea returned to Wichita and lived with his folks until August 1946. Bob took a couple of courses at Friends University, worked with his father in the family business, and utilized his aviation training as an instructor in the ground school two of his friends had recently established. Civilian flying proved to be both expensive and dull. "Readjustment" consisted of swapping a pair of Navy green pants for a pair of Army pinks and removing the insignia from the blues, which became his everyday suit for the next few years. Phyl, who had already taken up knitting, gave birth to their daughter Pamela on April 13, 1946. The boys were home, but they were no longer boys.

In the fall of 1946, Bob entered graduate school at Indiana University. The GI Bill provided support, and the Navy had taught men discipline—for which most of that academic generation has since been grateful. In 1950, having completed his work in English History under the wise and forbearing tutelage of Professor John J. Murray, Dr. Robert R. Rea secured a position as Assistant Professor of History at Alabama Polytechnic Institute, now Auburn University. His teaching career has centered in that institution, where he is Professor and a Distinguished Graduate Faculty Lecturer. Perhaps the correspondence of the war years contributed something to his production of a number of scholarly books and articles.

The world has changed much since 1943, though men very little. Jet fighters may measure their speed in terms of "machs" rather than knots, and computerized missiles may have replaced tracer bullets, but people seem strangely unaffected by the nuclear umbrella under which they live. Perhaps the assumption of security is necessary for our day-to-day survival. Perhaps we have also reached that point at which history cannot "repeat itself"—not that it ever does—nor would we wish it for our grandsons. Happily, there is much more to life than winning your wings of gold.

# APPENDIX:
# FLIGHT MANEUVERS MENTIONED

## 1. Glide
Controlled flight without the use of engine power, necessarily descending at greater than stalling speed.

## 2. Stall
Condition or speed at which level flight cannot be maintained. The nose of the plane drops (falls through the line of the horizon). If lateral equilibrium is not maintained, the plane will start into a spin (fall off on one wing).

## 3. Spin
Nose-first rotation, around the longitudinal axis, of a stalled and falling plane. In a normal spin, the plane rotates so that the pilot is on the inside of the rotation; in an outside spin, the pilot is on the outside of the aircraft's plane of rotation. Although spins were extremely dangerous in the early days of aviation, subsequent perfection of recovery technique enabled pilots to control the duration of rotation very precisely. A pilot who failed to recover from a spin and crashed into the ground was said to have "spun in."

## 4. Eights
Training maneuvers in which the symmetry of the figure 8 is utilized as a pattern to inculcate smooth, coordinated handling of a plane. The lazy 8 was performed while gliding, the nose of the plane being made to follow the pattern of a horizontal 8 by being brought, equidistant, above and below the horizon line while turning 45° to either side of the original line of flight. The pylon 8 was flown as a pattern around two ground markers, each centered in one half of the figure 8—a useful exercise preparatory to landing practice, as it requires careful coordination of all flight controls and attention to wind effect.

## 5. Skid
The effect of turning without sufficient banking (use of aileron), similar to the automobile skid. Undesirable in normal circumstances and dangerous when landing.

## 6. Slip

The process, when gliding, of increasing the rate of descent, without increasing speed, while maintaining directional control. The pilot angles the plane to the desired flight path and lowers (reduces the lift on) one wing. Used in precision landing to avoid overshooting the target area; also used in crosswind landing in light planes.

## 7. Falling Leaf

A maneuver performed in a stalled condition wherein the plane is repeatedly allowed to start a spin to one side, then forced into the start of a spin in the opposite direction, with resulting oscillation (hence the name) of 45° to either side of the initial line of flight.

## 8. Wingover

From normal flight attitude, the nose of the aircraft is pulled up until stalling speed is reached, at which point the aircraft is allowed to fall off into a 180° turn, flying speed being regained as the nose drops.

## 9. Loop

From normal flight attitude, the plane is flown upward and through a vertical circle, maintaining the line of flight and returning to the original position and attitude. An outside loop is flown by nosing down and flying through a circle, the pilot on the outside rather than the inside of the perimeter.

## 10. Split S

From normal flight attitude, the plane is turned on its back (upside down) by a half snap roll and then pulled through a half loop to normal flight position, thereby reversing the direction of flight with loss of altitude.

## 11. Immelmann

A maneuver named after the German ace Max Immelmann (d. 1916) in which the plane, at the top of a loop (when inverted) is rolled over to normal flight attitude, thereby accomplishing a 180° change of direction with maximum gain of altitude.

## 12. Aileron or Barrel Roll

A 360° rolling maneuver around the longitudinal axis of the plane, performed without use of the rudder. Direction of flight remains constant, but the aircraft moves around and at some distance from the line of flight, as around the outside of a barrel lying on its side.

## 13. Slow Roll

A 360° rolling maneuver along the line of flight, performed by coordination of aileron and rudder (in contrast to the barrel roll).

## 14. Snap Roll

A 360° rolling maneuver, maintaining the line of flight, while in a stalling condition—virtually a horizontal spin.

## 15. Lufbery Circle

A defensive tactic named after the famous World War I ace Raoul Lufbery, wherein a group of aircraft fly in a close nose-to-tail circle, thereby making it difficult for an attacking plane to gain a firing position directly behind any plane in the circle. The Lufbery circle provides no defense against deflection shooting.

## 16. Thach Weave

Tactic developed by Commander James Thach, USN, to counter the superior maneuverability of the Japanese Zero. Flying in a two-plane section, lead and wingman maintain sufficient lateral separation to be able to watch each other's tail and to turn in toward each other if threatened, thereby giving one plane a head-on shot at the attacker.

# NOTES

## Introduction

1. Clarke Van Vleet and William J. Armstrong, *United States Naval Aviation, 1910–1980* (Washington, D.C.: U.S. Government Printing Office, 1981), pp. 2–3; C. R. Roseberry, *Glenn Curtiss, Pioneer of Flight* (Garden City, N.Y.: Doubleday, 1972), pp. 294–95, 311.

2. Theodore Roscoe, *On the Seas and in the Skies* (New York: Hawthorne Books, 1970), pp. 29–33; Ellyson to Chambers, March 31, 1911, in Louis S. Casey, *Curtiss, the Hammondsport Years, 1907–1915* (New York: Crown, 1981), pp. 227–30.

3. Casey, *Curtiss*, pp. 227–30.

4. Roseberry, *Curtiss*, pp. 313, 317. See also George Van Deurs, *Anchors in the Sky: Spuds Ellyson, the First Naval Aviator* (San Rafael, Calif.: Presidio Press, 1978), and George Van Deurs, *Wings for the Fleet: A Narrative of Naval Aviation's Early Development, 1910–1916* (Annapolis: U.S. Naval Institute, 1966), pp. 33–35.

5. Van Vleet and Armstrong, *Naval Aviation*, pp. 4–8; Van Deurs, *Wings for the Fleet*, pp. 41–48, 54–55, 75–82.

6. Van Vleet and Armstrong, *Naval Aviation*, pp. 9–10; Report of Lt.-Comdr. H. C. Mustin to Commander United States Atlantic Fleet, May 19, 1914, copy in possession of Newton, provided in 1962 by Adrian O. Van Wyen, Historian, Office of the Deputy Chief for Naval Operations (Air). On the occupation of Veracruz, see Robert E. Quirk, *An Affair of Honor* (Lexington: University of Kentucky Press, 1962); Jack Sweetman, *The Landing at Veracruz, 1914* (Annapolis: U.S. Naval Institute, 1968).

7. Van Vleet and Armstrong, *Naval Aviation*, pp. 9–10. The death of Billingsley, who was tossed from a plane in the air, led to the Navy's adoption of seatbelts; Roscoe, *On the Seas*, p. 41.

8. Roscoe, *On the Seas*, p. 52.

9. Ibid., pp. 52–53; Van Vleet and Armstrong, *Naval Aviation*, pp. 13–15.

10. Francis K. Mason, *Aces of the Air* (New York: Mayflower Books, 1981), pp. 11, 23; Clark G. Reynolds, *The Carrier War* (Alexandria, Va.: Time-Life Books, 1982), pp. 10–13, 24–31.

11. E. B. Potter, *The Naval Academy Illustrated History of the United States Navy* (New York: Crowell, 1971), pp. 136–45; Van Vleet and Armstrong, *Naval Aviation*, pp. 24–35; Roscoe, *On the Seas*, pp. 63–67, 82–112; Archibald D. Turnbull and Clifford L. Lord, *History of United States Naval Aviation* (1949, reprint, New York: Arno Press, 1972), pp. 119–41.

12. Van Vleet and Armstrong, *Naval Aviation*, p. 23; Turnbull and Lord, *U.S. Aviation*, pp. 100–104, 126; Roscoe, *On the Seas*, pp. 69–70; Charles E. Matthews, "History of U.S. Naval Aviation during the World War: Training in America" (Record Group 45, National Archives and Records Service, Washington, D.C.), pp. 101–64.

13. George F. Pearce, *The U.S. Navy in Pensacola* (Pensacola: University Presses of Florida, 1980), pp. 152–54; Matthews, "Naval Aviation during the World War," pp. 18–21.

14. Pearce, *Pensacola*, pp. 154–55. Authorized flight clothing and the working uniform of a canvas, khaki, or moleskin coverall is described in Van Vleet and Armstrong, *Naval Aviation*, pp. 25–27.

15. Pearce, *Pensacola*, p. 157; Matthews, "Naval Aviation during the World War," pp. 155–56.

16. Charles M. Melhorn, *Two Block Fox: The Rise of the Aircraft Carrier, 1911–1929* (Annapolis: U.S. Naval Institute Press, 1974), pp. 56–57, 74–110, 112; Alfred F. Hurley, *Billy Mitchell: Crusader for Air Power* (New York: Watts, 1964), pp. 67, 97, 107; Clark G. Reynolds, *The Fast Carriers: The Forging of an Air Navy* (New York: McGraw-Hill, 1968), pp. 14–15.

17. Hurley, *Mitchell*, pp. 90–107; Turnbull and Lord, *U.S. Aviation*, p. 283.

18. Van Vleet and Armstrong, *Naval Aviation*, p. 53.

19. Peter C. Smith, *The History of Dive Bombing* (Annapolis: Nautical & Aeronautical Publishing, 1981), pp. 13–30.

20. For the evolution of designs, see the photographic sections of Turnbull and Lord, *U.S. Aviation*, and Van Vleet and Armstrong, *Naval Aviation*. The 1929 and 1930 fleet exercises are described in Turnbull and Lord, *U.S. Aviation*, pp. 270–72.

21. Pearce, *Pensacola*, pp. 162, 164. On enlisted Aviation Pilots, see W. E. Scarborough, "Wings for the Fleet," *Air Classics* 7 (October 1970): 32–41, 57–58; George Van Deurs, "Navy Wings between the Wars" (Van Deurs Papers, Naval Historical Foundation, Washington, D.C.), p. 81.

22. Pearce, *Pensacola*, pp. 164–65, 172–74.

23. D. C. Ramsey, "The Course of Training for Student Naval Aviators," U.S. Naval Institute *Proceedings* 53 (1927): 303; Barrett Studley, "Flight Training of Student Naval Aviators," ibid., p. 765.

24. Studley, "Flight Training," pp. 764–65.

25. The foregoing descriptions of flight instruction are from Studley, "Flight Training," pp. 765–72. See also Pearce, *Pensacola*, pp. 166–67; Van Deurs, "Between the Wars," pp. 89–102.

26. Studley, "Flight Training," pp. 768–69. See also Pearce, *Pensacola*, pp. 167–68, and Turnbull and Lord, *U.S. Aviation*, p. 268.

27. Studley, "Flight Training," p. 764; Ramsey, "Course of Training," pp. 303, 305–6.

28. "U.S. Naval Aviation Accident Statistics and Flying Hours," U.S. Naval Institute *Proceedings* 56 (1930): 941; Ramsey, "Course of Training," pp. 304, 306–7; Pearce, *Pensacola*, pp. 165–66.

29. American interest in aviation is compared to religious fervor by Joseph Corn, *The Winged Gospel* (Palo Alto: Stanford University Press, 1983).

30. Turnbull and Lord, *U.S. Aviation*, pp. 230–32; Van Vleet and Armstrong, *Naval Aviation*, p. 64; Allen E. Morgan, "Naval Reserve Aviation Base, Long Beach, 1929–1942," *Journal of the American Aviation Historical Society* 16 (1971): 280–84.

31. Ramsey, "Course of Training," p. 303; D. C. Ramsey, "Aviation Training for Midshipmen and Line Officers," U.S. Naval Institute *Proceedings* 57 (1931): 18–21; Van Deurs, "Between the Wars," p. 83, takes a less sanguine view of the program at Annapolis.

32. "Aeronautical Organization of U.S. Navy, Fiscal Year, 1931," U.S. Naval Institute *Proceedings* 56 (1930): 934–36, 944.

33. "Syllabus for the Training of Student Naval Aviators and Student Naval Aviation Pilots," U.S. Naval Institute *Proceedings* 56 (1930): 937–41; Van Vleet and Armstrong, *Naval Aviation*, pp. 79–80.

34. Van Vleet and Armstrong, *Naval Aviation*, pp. 77, 83, 85.

35. Ibid., p. 84.

36. 49 Stat. 156 (1935); Turnbull and Lord, *U.S. Aviation*, pp. 291–95; Joe Hill, *Some Early Birds: The Memoirs of a Naval Aviation Cadet, 1935–1945* (Manhattan, Kans.: Sunflower University Press, 1983), p. 2.

37. Van Vleet and Armstrong, *Naval Aviation*, pp. 86–88.

38. Hill, *Early Birds*, pp. 1–12.

39. Ibid., p. 14.

40. Ibid., pp. 14–15; "Indoctrinal Training Syllabus," ibid., pp. 15–18.

41. Ibid., p. 43.

42. Ibid., pp. 37–38.

43. Ibid., pp. 21, 23, 25–26, 31, 34–35; Roscoe, *On the Seas*, pp. 242–43.

44. Hill, *Early Birds*, pp. 30–32, does not mention aerology in the ground school curriculum, apparently an oversight.

45. Ibid., 40–42.

46. John R. Poppen, "Why Am I Not Fit to Fly," U.S. Naval Institute *Proceedings* 62 (1936): 811–16.

47. Hill, *Early Birds*, pp. 27, 30, 39.

48. 53 Stat. 819 (1939); Van Vleet and Armstrong, *Naval Aviation*, p. 93.

49. Van Vleet and Armstrong, *Naval Aviation*, pp. 91–94; Van Deurs, "Between the Wars," pp. 125–26.

50. G. M. Fennemore, M. B. Chambers, and A. F. Vaupel, "Aspects of Aviation Training, 1940–45," vol. 14, U.S. Office of Naval Operations (Air), pp. 3–5.

51. Van Vleet and Armstrong, *Naval Aviation*, p. 103.

52. Ibid., p. 101.

53. Fennemore, Chambers, and Vaupel, "Aviation Training," pp. 11–19, 25–27.

54. Ibid., 28–44.

55. By 1943, these Air Stations included Anacostia, Md.; Atlanta, Ga.; Dallas, Tex.; Grosse Ile, Mich.; Hutchinson, Kans.; Los Alamitos, Calif.; Minneapolis, Minn.; New Orleans, La.; Olathe, Kans.; Ottumwa, Iowa; Pasco, Wash.; Peru, Ind.; Squantum, Mass.; St. Louis, Mo.; Willow Grove, Pa.; and Memphis, Tenn.

56. BuNav *Bulletin*, no. 300 (March 1, 1942): 17–18.

57. Navy Dept. press releases, December 17 and 21, 1942.

58. Patricia Strickland, *The Putt-Putt Air Force* (Washington, D.C.: U.S. Government Printing Office, 1975), pp. 1–6; A. O. Van Wyen, "The Civil Aeronautics Authority War Training Service," vol. 15, U.S. Office of Naval Operations (Air), pp. 18–28, 30–32, 51–72.

59. The *Fledgling* (Wm. Jewell NFPS), August 13, 1943, p. 3.

60. 56 Stat. 737 (1942); Navy Dept. press release, April 22, 1942; BuPers *Information Bulletin*, no. 334 (January 1945): 73; A. Russell Buchanan, *The Navy's Air War* (New York: Harper & Bros., 1946), vol. 1, pp. 308–9.

61. Rear Admiral L. E. Denfeld to Chiefs of Air Training, June 23, 1944, and Chief of Naval Personnel to Chief of Naval Air Training, June 17, 1944; Pers 400, A21/P11-1; BuPers *Information Bulletin*, no. 328 (July 1944): 69; "Aviation Personnel, 1939–1945," vol. 22, U.S. Office of Naval Operations (Air), pp. 217–51.

62. BuPers *Information Bulletin*, no. 334 (January 1945): 73–74.

63. *Liberty* [Mo.] *Tribune*, December 3, 1942, p. 1; the *Student* (Wm. Jewell), December 7, 1942, p. 1; the *Fledgling*, September 17, 1943, pp. 1–2; *Kansas City Star*, December 18, 1942, p. 4; William Jewell College *Bulletin*, vol. 36, no. 2 (December 1942), pp. 1, 4; *Report of the President of William Jewell College to the Board of Trustees*, May 15, 1943, pp. 13–14, 18–19.

64. *Liberty Tribune*, January 7, 1943, p. 1.

65. Ibid., December 31, 1942, to June 3, 1943; *Liberty Advance*, January 25, 1943, to April 26, 1943; the *Fledgling*, September 17, 1943, pp. 1–2.

66. *Liberty Tribune*, December 31, 1942, p. 1; February 11, 1943, p. 1; April 8, 1943, pp. 1, 5; September 9, 1943, p. 8; the *Fledgling*, September 17, 1943, p. 1.

67. The *Fledgling*, June 25, 1943, p. 1; July 30, 1943; *Liberty Tribune*, June 10, 1943, p. 1; *Liberty Advance*, August 30, 1943, p. 1.

68. These activities are described in text and pictures in the *Fledgling*, June 25, July 30, August 13, September 17, 1943.

69. The *Fledgling*, September 17, 1943, p. 1; *Liberty Tribune*, September 28, 1944, p. 1.

70. Strickland, *Putt-Putt Air Force*, p. 6; Robert Taft, *The Years on Mount*

*Oread* (Lawrence: University Press of Kansas, 1955), pp. 153–71; Virginia Adams, *On the Hill: A Pictorial History of the University of Kansas* (Lawrence: University Press of Kansas, 1983), pp. 126–39; Clifford S. Griffin, *The University of Kansas: A History* (Lawrence: University Press of Kansas, 1974), pp. 479–500.

71. Van Wyen, "War Training Service," pp. 44–47.

72. Larry Perl, *Calm and Serene on Thy Hill: A Retrospective of the University of Iowa* (Iowa City: University Alumni Association, 1978), pp. 170–97.

73. Ryland Wesley Crary, "History of the State University of Iowa: The Liberal Arts College in the Gilmore and Hancher Administrations" (Ph.D. diss., University of Iowa, 1946), pp. 207–9.

74. Ibid., pp. 209–14; Perl, *Calm and Serene*, pp. 188–92, 194, 207–8.

75. Denfeld to Commandant, Ninth Naval District, February 17, 1942, and Football Program, November 13, 1943, in "University of Iowa," V-5, Box 7, Naval Historical Center, Operational Archives Branch, Washington, D.C.

76. Interviews of Robert R. Rea by Wesley Phillips Newton, 1984–1985.

77. "Olathe, Kan.," in Paolo E. Coletta and K. Jack Bauer, eds., *United States Navy and Marine Corps Bases, Domestic* (Westport, Conn.: Greenwood Press, 1985), pp. 406–7; "History of U.S. Naval Air Station, Olathe, Kansas," Naval Historical Center, Operational Archives Branch, Washington, D.C., pp. 3–5.

78. "NAS Olathe," pp. 1, 10–13.

79. Bureau of Aeronautics *News Letter* 18, no. 175 (August 1, 1942): 16; Rea interviews.

80. "NAS Olathe," p. 8; Rea interviews.

81. "NAS Olathe," pp. 7–8; Rea interviews.

82. Rea memorandum no. 1, in possession of W. P. Newton.

83. George F. Pearce, "Pensacola, Fla.," in Coletta and Bauer, *U.S. Navy Bases*, pp. 471–72; "Naval Air Training Pensacola," *Naval Aviation News*, no. 212 (March 1, 1944): 27; "Thousands of Students Receive Their Navy Wings at Pensacola," ibid., p. 28; Van Vleet and Armstrong, *Naval Aviation*, p. 128; "Flight Surgeons Pilot Planes," *Naval Aviation News*, no. 212 (March 1, 1944): 31.

84. Frank Robinson, *The British Pensacola Battalion* (Manhattan, Kans.: Sunflower University Press, 1983), pp. 1, 18; "Thousands of Students," p. 28.

85. "Corpus Christi, Tex.," in Coletta and Bauer, *U.S. Navy Bases*, pp. 153–54; "Corpus Is University of the Air," *Naval Aviation News*, no. 218 (June 1, 1944): 28.

86. "Corpus Christi," in Coletta and Bauer, *U.S. Navy Bases*, p. 154; "Cadets Get Tough Intermediate Training, Win Wings at Corpus," "Corpus Christi Covers Huge Area," *Naval Aviation News*, no. 218 (June 1, 1944): 29–31; "Instructors Fit Ground Training to Specific Types of Planes," ibid., p. 30; Rea interviews.

87. "Strict Athletic Program Keeps Aviation Cadet in Top Form," *Naval*

*Aviation News,* no. 218 (June 1, 1944): 32; "Thousands of Students," ibid., p. 28; "History of Physical and Military Training Program at Corpus Christi, Texas," Naval Historical Center, Operational Archives Branch, Washington, D.C.

88. "The Gunairstructor," *Naval Aviation News,* no. 204 (November 1, 1943): 1; Rea interviews.

89. Fennemore, Chambers, and Vaupel, "Aviation Training," pp. 160–73; Rea interviews.

90. Rea memorandum no. 2, 1984.

91. Rea interviews.

92. Rea memorandum no. 3, 1984.

93. Bureau of Aeronautics *News Letter,* no. 175 (August 1942): 36.

94. Rea interviews.

95. "Summary and Description of U.S. Naval Aviation Program, August 10, 1943," pp. 5–7, Office of Chief of Naval Operations, Naval Aviation History and Archives, Washington, D.C.; Bureau of Aeronautics *News Letter,* no. 177 (September 1, 1942): 34; Buchanan, *Navy's Air War,* vol. 1, pp. 312–14.

96. "Green Cove Springs, Fla.," in Coletta and Bauer, *U.S. Navy Bases,* pp. 221–22; "Summary and Description," p. 6.

97. Rea interviews.

98. "Los Alamitos, Calif.," in Coletta and Bauer, *U.S. Navy Bases,* pp. 291–93.

99. Rea interviews.

## Chapter 2

1. A. J. P. Taylor, *The Second World War* (New York: Perigee, 1983), p. 139. Taylor included all four Japanese carriers sunk at Midway in the "five minutes" work of the SBDs. The fourth victim was actually struck some hours later. See John Costello, *The Pacific War, 1941–1945* (New York: Quill, 1982), pp. 245–63.

2. A. Russell Buchanan, *The United States and World War II* (New York: Harper & Row, 1964), vol. 1, pp. 229–40.

3. Costello, *Pacific War,* pp. 324–34, 360–73.

4. C. L. Sulzberger, *World War II* (New York: McGraw-Hill, 1970), pp. 130–31.

5. Martin Middlebrook, *The Battle of Hamburg* (New York: Scribner's, 1980), pp. 8–12; James L. Stokesbury, *A Short History of World War II* (New York: Quill, 1980), pp. 277–80; Wilbur H. Morrison, *Fortress without a Roof* (New York: St. Martin's, 1982), pp. 29–36, 51–59.

6. Stokesbury, *Short History,* pp. 128–31, 235–41; Taylor, *Second World War,* pp. 97–118, 141–47; Buchanan, *U.S. & World War II,* vol. 1, pp. 115–16.

7. Richard Polenberg, *War and Society* (Philadelphia: Lippincott, 1972), pp. 188–89.

8. Both Newton and Rea recall this attitude appearing between the fall of 1942 and the spring of 1943, when both were inducted into service. Rea wrote to Phyllis Edwards, December 13, 1942, "The general atmosphere at school is hardly what you'd call Christmas-like. With the men in school being classified 1-A . . . , others leaving, everything has a rather desperate 'last stand' air about it."

9. John Morton Blum, *V Was for Victory* (New York: Harcourt Brace Jovanovich, 1976), p. 24.

10. For the movies described in this paragraph, see advertisements in the *Wichita Eagle* and *Beacon*. They are characterized in Steven H. Scheuer, ed., *Movies on TV, 1985* (Toronto: Bantam, 1985).

11. *New York Times,* November 29, 30, 1942.

12. The era of the Big Bands is described in George T. Simon, *The Big Bands,* 4th ed. (New York: Schirmer, 1981), and in his *Glenn Miller and His Orchestra* (New York: Crowell, 1974).

13. "The Middle West," *Life,* November 9, 1942, pp. 103–9.

14. Blum, *Victory,* pp. 226–27.

15. "The Middle West," p. 107.

16. Buchanan, *U.S. & World War II,* vol. 1, p. 118; Stokesbury, *Short History,* pp. 241–42; Taylor, *Second World War,* p. 128.

17. Winston S. Churchill, *The Second World War* (Boston: Houghton Mifflin, 1953), vol. 4, *The Hinge of Fate,* pp. 678–80; vol. 5, *Closing the Ring,* p. 520; Stokesbury, *Short History,* pp. 280–82.

18. Polenberg, *War and Society,* pp. 161–67; Blum, *Victory,* p. 241.

19. Noel Busch, "The King of Arabia," *Life,* May 31, 1943, pp. 76–77; Stephen S. Wise, "In Reply to King Ibn Saud," *Life,* June 21, 1943, pp. 37–38; Lessing J. Rosenwald, "Reply to Zionism," *Life,* June 28, 1943, p. 11.

20. Costello, *Pacific War,* pp. 403–4.

21. Morrison, *Fortress,* pp. 136–39, 147–53.

22. The phrase "nightmare fighters" is from Randall Jarrell's "The Death of the Ball Turret Gunner," in his *Little Friend, Little Friend* (New York: Dial, 1945), p. 58.

23. Buchanan, *U.S. & World War II,* vol. 1, pp. 166–72; Taylor, *Second World War,* pp. 173–74.

24. "The Cruise of the 'Campbell,' " *Life,* July 5, 1943, pp. 57–73; "Race Riot in Detroit," ibid., pp. 93–102; "Harlem's Wild Rampage," *Life,* August 16, 1943, pp. 32–33.

## Chapter 3

1. Buchanan, *U.S. & World War II,* vol. 2, pp. 302–4; Stokesbury, *Short History,* pp. 297–99.

2. From Martin Caidin, *Black Thursday* (New York: Bantam, 1981).

3. Morrison, *Fortress,* chap. 17; Martin Middlebrook, *The Schweinfurt-Regensburg Mission* (New York: Scribner's, 1983), chap. 20.

4. Costello, *Pacific War,* pp. 428–39; Roscoe, *On the Seas,* pp. 388–89.

5. "The Furnaces Roar," *Life,* September 6, 1943, pp. 23–30.

6. "One of Our Bombers Is Missing," "Leotards," *Life,* September 13, 1943, pp. 41–42, 44, 47–48, 50, 52, 54.

## Chapter 4

1. John Scott, "Bombing of Berlin," *Life,* December 13, 1943, p. 16.

2. "1944," "U.S. Civilians Buy Their First Jeeps," *Life,* January 3, 1944, pp. 20, 24.

3. Most recently Rea has published *George Gauld: Surveyor and Cartographer of the Gulf Coast* (Gainesville, Fla.: University Presses of Florida, 1982), with John D. Ware; *The Log of H.M.S. Mentor, 1780–1781* (Pensacola: University Presses of Florida, 1982), with James A. Servies; and "Master James Cook and Gulf Coast Cartography," *Florida Historical Quarterly* 63 (1985): 280–302.

## Chapter 5

1. "Pre-Invasion," *Life,* March 6, 1944, pp. 25–33.

2. Costello, *Pacific War,* pp. 449–53.

3. "Life Visits with the Hatfields and the McCoys," *Life,* May 22, 1944, p. 115.

## Chapter 6

1. Costello, *Pacific War,* pp. 478–83; Roscoe, *On The Seas,* p. 405; Thomas M. Coffey, *Hap* (New York: Viking, 1982), pp. 336–40.

2. Costello, *Pacific War,* pp. 501, 506, 517–18, 564–65, 574–78.

## Chapter 7

1. Stokesbury, *Short History,* pp. 286, 351–63.

2. Costello, *Pacific War,* pp. 525–30, 542–66.

# BIBLIOGRAPHY

## General Background

Blum, John Morton. *V Was for Victory*. New York: Harcourt Brace Jovanovich, 1976.

Buchanan, A. Russell. *The United States and World War II*. New York: Harper & Row, 1964.

Caidin, Martin. *Black Thursday*. New York: Bantam, 1981.

Casey, Louis S. *Curtiss, the Hammondsport Years, 1907–1915*. New York: Crown, 1981.

Churchill, Winston S. *The Second World War*. Boston: Houghton Mifflin, 1953.

Coffey, Thomas M. *Hap*. New York: Viking, 1982.

Corn, Joseph. *The Winged Gospel*. Palo Alto: Stanford University Press, 1983.

Costello, John. *The Pacific War, 1941–1945*. New York: Quill, 1982.

Hurley, Alfred F. *Billy Mitchell: Crusader for Air Power*. New York: Watts, 1964.

Middlebrook, Martin. *The Battle of Hamburg*. New York: Scribner's, 1980.

———. *The Schweinfurt-Regensburg Mission*. New York: Scribner's, 1983.

Morrison, Wilbur H. *Fortress without a Roof*. New York: St. Martin's, 1982.

Polenberg, Richard. *War and Society*. Philadelphia: Lippincott, 1972.

Quirk, Robert E. *An Affair of Honor*. Lexington: University of Kentucky Press, 1962.

Roseberry, C. R. *Glenn Curtiss, Pioneer of Flight*. Garden City, N.Y.: Doubleday, 1972.

Scheuer, Steven H., ed. *Movies on TV, 1985*. Toronto: Bantam, 1985.

Simon, George T. *The Big Bands*. 4th ed. New York: Schirmer, 1981.

———. *Glenn Miller and His Orchestra*. New York: Crowell, 1974.

Stokesbury, James L. *A Short History of World War II*. New York: Quill, 1980.

Sulzberger, C. L. *World War II*. New York: McGraw-Hill, 1970.

Sweetman, Jack. *The Landing at Veracruz, 1914*. Annapolis: U.S. Naval Institute, 1968.

Taylor, A. J. P. *The Second World War*. New York: Perigee, 1983.

## History of Naval Aviation

Buchanan, A. Russell. *The Navy's Air War*. New York: Harper & Bros., 1946.

Caidin, Martin. *Golden Wings*. New York: Arno, 1974.

Coletta, Paolo E., and Bauer, K. Jack., eds. *United States Navy and Marine Corps Bases, Domestic*. Westport, Conn.: Greenwood Press, 1985.

Hill, Joe. *Some Early Birds. The Memoirs of a Naval Aviation Cadet, 1935–1945.* Manhattan, Kans.: Sunflower University Press, 1983.

Mason, Francis K. *Aces of the Air.* New York: Mayflower Books, 1981.

Melhorn, Charles M. *Two Block Fox: The Rise of the Aircraft Carrier, 1911–1929.* Annapolis: U.S. Naval Institute Press, 1974.

Morgan, Allen E. "Naval Reserve Aviation Base, Long Beach, 1929–1942," *Journal of the American Aviation Historical Society* 16 (1971).

Owsley, Frank L., Jr., and Wesley Phillips Newton, "Eyes in the Skies" (Naval Aviation at Vera Cruz, 1914), Supplement/April 1986, U.S. Naval Institute *Proceedings,* 17–25.

Pearce, George F. *The U.S. Navy in Pensacola.* Pensacola: University Presses of Florida, 1980.

———. "Pensacola Naval Air Station, 1914–1986," *Pensacola History Illustrated* 2 (Spring 1986): 2–9.

Potter, E. B. *The Naval Academy Illustrated History of the United States Navy.* New York: Crowell, 1971.

Reynolds, Clark G. *The Carrier War.* Alexandria, Va.: Time-Life Books, 1982.

———. *The Fast Carriers: The Forging of an Air Navy.* New York: McGraw-Hill, 1968.

Robinson, Frank. *The British Pensacola Battalion.* Manhattan, Kans.: Sunflower University Press, 1983.

Roscoe, Theodore. *On the Seas and in the Skies.* New York: Hawthorne Books, 1970.

Russell, Sandy, ed. *Naval Aviation, 1911–1981: A Pictorial Study.* Washington, D.C.: U.S. Government Printing Office, 1984.

Scarborough, W. E. "Wings for the Fleet," *Air Classics* 7 (October 1970).

Smith, Peter C. *The History of Dive Bombing.* Annapolis: Nautical & Aeronautical Publishing, 1981.

Strickland, Patricia. *The Putt-Putt Air Force.* Washington, D.C.: U.S. Government Printing Office, 1975.

Turnbull, Archibald D., and Lord, Clifford L. *History of United States Naval Aviation.* 1949. Reprint. New York: Arno Press, 1972.

Van Deurs, George. *Anchors in the Sky: Spuds Ellyson, the First Naval Aviator.* San Rafael, Calif.: Presidio Press, 1978.

———. *Wings for the Fleet: A Narrative of Naval Aviation's Early Development, 1910–1916.* Annapolis: U.S. Naval Institute Press, 1966.

Van Vleet, Clarke, and Armstrong, William J. *United States Naval Aviation, 1910–1980.* Washington, D.C.: U.S. Government Printing Office, 1981.

Weitzman, Ellis, and Bedell, Ralph C. "The Central Examining Board for the Training of Naval Air Cadets," *Psychological Bulletin* 41 (1944).

Weitzman, Ellis, and McNamara, Walter J. "Techniques Used in Analyzing the Learning Achievements of Naval Aviation Cadets," *Journal of Educational Psychology* 35 (1944).

## Institutional Histories

Adams, Virginia. *On the Hill: A Pictorial History of the University of Kansas.* Lawrence: University Press of Kansas, 1983.

Crary, Ryland Wesley. "History of the State University of Iowa: The Liberal Arts College in the Gilmore and Hancher Administrations." Ph.D. diss., University of Iowa, 1946.

Griffin, Clifford S. *The University of Kansas: A History.* Lawrence: University Press of Kansas, 1974.

Perl, Larry. *Calm and Serene on Thy Hill: A Retrospective of the University of Iowa.* Iowa City: University Alumni Association, 1978.

*Report of the President of William Jewell College to the Board of Trustees,* May 15, 1943.

Taft, Robert. *The Years on Mount Oread.* Lawrence: University Press of Kansas, 1955.

*William Jewell College Bulletin,* 1942.

## Newspapers and Periodicals

*Kansas City* [Mo.] *Star,* 1942–1943.
*Liberty* [Mo.] *Advance,* 1943.
*Liberty* [Mo.] *Tribune,* 1942–1943.
*Life,* 1942–1945.
Naval Flight Preparatory School, Liberty, Mo., the *Fledgling,* 1943.
*New York Times,* 1942.
U.S. Naval Institute *Proceedings.*
U.S. Navy, Bureau of Aeronautics *News Letter,* 1942.
———, Bureau of Navigation *Bulletin,* 1942–1945.
———, Bureau of Personnel *Information Bulletin,* 1944–1945.
———, *Naval Aviation News,* 1943–1944.
William Jewell College, Liberty, Mo., *Student,* 1942–1943.

## Unpublished Sources and Repositories

Matthews, Charles E. "History of U.S. Naval Aviation during the World War: Training in America." Record Group 45, National Archives and Records Service, Washington, D.C.

Naval Aviation History and Archives, Office of the Chief of Naval Operations, Washington, D.C.

Naval Historical Center, Operational Archives Branch, Washington, D.C.

Rea, Robert R., and Newton, Wesley P. Interviews and memoranda, Auburn, Ala., 1984–1985.

U.S. Office of Naval Operations (Air). Deputy Chief of Naval Operations (Air). Vol. 13, Aviation Training, 1911–1939; vol. 14, Aviation Training, 1940–1945; vol. 15, The Civil Aeronautics Authority War Training Service; vol. 21, Aviation Personnel; vol. 22, Aviation Personnel, 1939–1945.

Van Deurs, George. "Navy Wings between the Wars." Van Deurs Papers, Naval Historical Foundation, Washington, D.C.

# INDEX

Accidents: 1913–14, 3; 1914–18, 5–6; 1918–24, 7; 1920s, 10–11; 1935–41, 17; at Lawrence (WTS), 90; at Olathe (Primary), 33, 140, 144, 146, 148, 150, 152, 157, 164, 166; at Beeville (Instrument), 205; at Kingsville (Advanced), 228, 235, 237, 242; at Pre-operational, 250; at Green Cove Springs (Operational), 260, 264, 271–72; in Air Group 98, 289, 300–302, 304
Aeronautic Center, 3
Aircraft
    **Pre-World War II:** *Triad,* 1–2; Curtiss flying boat, 3; Curtiss N-9 Jenny, 5, 8–9; Curtiss R-type, 5; Consolidated NY-1, 9–10, 13; Curtiss CS-1, 10; Vought UO-1, 11; Ford Trimotor, 46;
    **light and training planes:** Piper Cub J-3, 27, 81, flight in, 83–97; Taylorcraft, 296; Navy N3N, 13, 16, 27, 90, 93, 96–97, 138–40, 224, flight in, 98–102; Stearman N2S, 13, 16, described, 33–34, 133–35, 138, flight in, 140–74; Vultee SNV Valiant, 38, 176, flight in, 183–89; North American SNJ Texan, 39–40, 43, 176, 197, 203–205, 220, flight in, 224–43, 254–55, 267, 283;
    **fighters:** F4B, 15; Grumman F4F Wildcat (FM), 19, 53, 299, 301–303; Grumman F6F Hellcat, 42, 45, 80, 245, 260, 294, 299; Grumman F7F, 275; F4U Corsair, 42–45, 132, 260–61, flight in, 268–87, 293–303; F4U-4, 303; FR1, 298; P-47, 255;
    **divebombers:** Douglas SBD Dauntless, 13, 19, 41, 53, 133, 176, 189, 243–44, flight in, 248–54, 271, 275; Curtiss SB2C Helldiver, 45, 133, 279, 299;
    **torpedo:** TBF Avenger, 45, 133, 291, 299;
    **scout:** OS2U Kingfisher, 39, 179–80;
    **multiengine:** Beech SNB, 39, 192, 271; PBY Catalina, 39, 153, 179–80, 195, 230, 255, 271; C-47, 271, 299; Lodestar, 146; Lockheed PV Ventura, 204; C-54 Skymaster, 154
Air Group 98, 44–45, 289–308
*Akron* (airship), 6
Albuquerque, New Mexico, 47
Alice, Texas, 177, 189
Alpha Kappa Tau (fraternity), 48, 64, 254, 302
Annapolis, 2–3. *See also* Naval Academy
Army Air Force, 4, 6, 15, 22, 53–54, 58, 60, 79–80, 132, 199, 213, 298
Auburn University, 15, 309

Barstad, Lieutenant (instructor), 250
Beeville, Texas, 35. *See also* Rea, Robert Right
Bellinger, Patrick N. L., 2–4
Bice, W. O., 24–25, 70
Billingsley, W. D., 3
Boston University, 51
Brow, H. J., 32
Brown, Ensign, 77
Brownsville, Texas, 241
Bureau of Aeronautics, 6, 11, 19
Bureau of Navigation, 19
Bureau of Personnel, 23
Burns, David M., 15

Cadet program: established (1935), 13; described (1935–39), 14–18, (1942–45), 21–23. *See also* Training programs; Rea, Robert Right
Carrier Air Support Unit, 44
Carrier landing, 43–45. *See also* Rea, Robert Right
Cecil Field, 279
Central Christian Church, 47, 214
Chamberlain, Clarence, 15
Chambers, Washington Irving, 1
*Charger,* 43

# ABOUT THE AUTHORS

**Wesley Phillips Newton** is Professor of History, Auburn University. He received his M.A. and Ph.D. degrees from The University of Alabama. Among his publications are *The Perilous Sky: United States Aviation Diplomacy and Latin America, 1919–1931; Delta: The History of an Airline* (with W. David Lewis). He is also contributing author and editor of *Air Force Combat Units of World War II,* volume I of the USAF lineage history, edited by Maurer Maurer.

**Robert R. Rea** is Professor of History, Auburn University. He received his M.A. and Ph.D. degrees from Indiana University. Among his publications are *The English Press in Politics, 1760–1774; To Prove a Villain* (with Taylor Littleton); *Memoire Justificatif of the Chevalier Montaut de Monberault* (with Milo B. Howard, Jr.); *The Minutes, Journals, and Acts of the General Assembly of British West Florida* (with Milo B. Howard, Jr.); *George Gauld: Surveyor and Cartographer of the Gulf Coast* (with John D. Ware); and *The Log of H.M.S. Mentor, 1780–1781* (with James A. Servies).